Contents

The artist for this issue is Doug Craft

CYBERNETICS & HUMAN KNOWING
A Journal of Second-Order Cybernetics,
Autopoiesis & Cyber-Semiotics
ISSN: 0907-0877

Cybernetics and Human Knowing is a quarterly international multi- and transdisciplinary journal focusing on second-order cybernetics and cybersemiotic approaches.

The journal is devoted to the new understandings of the self-organizing processes of information in human knowing that have arisen through the cybernetics of cybernetics, or second order cybernetics its relation and relevance to other interdisciplinary approaches such as C.S. Peirce's semiotics. This new development within the area of knowledge-directed processes is a non-disciplinary approach. Through the concept of self-reference it explores: cognition, communication and languaging in all of its manifestations; our understanding of organization and information in human, artificial and natural systems; and our understanding of understanding within the natural and social sciences, humanities, information and library science, and in social practices like design, education, org- anization, teaching, therapy, art, management and politics. Because of the interdisciplinary character articles are written in such a way that people from other domains can understand them. Articles from practitioners will be accepted in a special section. All articles are peer-reviewed.

Subscription Information

Price: Individual $104/£52. Institutional: $230/£115. 50% discount on full set of back volumes. Payment by check in $US or £UK, made payable to Imprint Academic to PO Box 200, Exeter EX5 5HY, UK, or Visa/Mastercard/Amex). sandra@imprint.co.uk

Editor in Chief: Søren Brier, Professor in semiotics at the Department of International Culture and Communication Studies attached to the Centre for Language, Cognition, and Mentality, Copenhagen Business School, Dalgas Have 15, DK-2000 Frederiksberg, Denmark, Tel: +45 38153246. sb.ikk@cbs.dk

Associate editor: Jeanette Bopry, Instructional Sciences, National Institute of Education, 1 Nanyang Walk, Singapore 637616. jeanette.bopry@gmail.com

Associate editor: Dr. Paul Cobley, Reader in Communications, London Metropolitan University, 31 Jewry Street, London EC3N 2EY. p.cobley@londonmet.ac.uk

Website editor: Argyris Arnellos, Department of Product and Systems Design Engineering of the University of the Aegean, Syros, Greece. arar@aegean.gr

Managing editor: Phillip Guddemi, The Union Institute and University, Sacramento CA, USA. pguddemi@well.com

Art editor and ASC-column editor: Pille Bunnell, Royal Roads University, Victoria BC, Canada. pille@interchange.ubc.ca

Journal homepage: www.chkjournal.org
Full text: www.ingenta.com/journals/browse/imp

Book Reviews: Publishers are invited to submit books for review to the Editor.

Instructions to Authors: To facilitate editorial work and to enhance the uniformity of presentation, authors are requested to send a file of the paper to the Editor on e-mail. If the paper is accepted after refereeing then to prepare the contribution in accordance with the stylesheet information at www.chkjournal.org

Manuscripts will not be returned except for editorial reasons. The language of publication is English. The following information should be provided on the first page: the title, the author's name and full address, a title not exceeding 40 characters including spaces and a summary/ abstract in English not exceeding 200 words. Please use italics for emphasis, quotations, etc. Email to: sbr.lpf@cbs.dk

Drawings. Drawings, graphs, figures and tables must be reproducible originals. They should be presented on separate sheets. Authors will be charged if illustrations have to be re-drawn.

Style. CHK has selected the style of the APA (*Publication Manual of the American Psychological Association*, 5th edition) because this style is commonly used by social scientists, cognitive scientists, and educators. The APA website contains information about the correct citation of electronic sources. The APA Publication Manual is available from booksellers. The Editors reserve the right to correct, or to have corrected, non-native English prose, but the authors should not expect this service. The journal has adopted U.S.English usage as its norm (this does not apply to other native users of English). For full APA style informations see: apastyle.apa.org

Accepted WP systems: MS Word and rtf.

Craft, D. (2009). *Simulacra and Simulation*. Microphoto: Ascorbic Acid Melt Crystals; 34 x 21 cm.

Cybernetics and Human Knowing. Vol. 16, nos. 3-4, pp. 5-10

Foreword:
The Transdisciplinary Contribution of Semiotics to Cybersemiotics

Søren Brier, Phillip Guddemi, Pille Bunnell, and Jeanette Bopry

The guiding idea behind the present thematic double issue is a presentation of the transdisciplinary scope of the new semiotics offering a deeper and broader framework than the structuralist semiology that has been the foundation of most European semiotic analyses of culture, texts and languages.

The once fruitful paradigm based on Saussure's semiological linguistic seems to have run out of steam during the last twenty years. Among the reasons are: (1) Structuralism has lost its former acuity by being used too broadly in cultural studies without keeping its structuralist framework clear—as well as its ontological assumptions and methodological limitations—and this is reflected in its methods. (2) It has limited itself in its theoretical framework to language as an autonomous system and then to culture as a system whose structure depends on the structure of the language. (3) Structuralism lacks a theory of consciousness and of the subject's agency in the production of meaning. Actually, structuralism and semiology have their goal—scientifically at least—to get rid of the roie of the subject in the production of meaning and as an important agency for the causality of social and cultural action. (4) Historically, structuralism focuses on breaks in long-lasting epistemes (Foucault) when it reluctantly adopts a diachronic perspective. Originally, it was conceived without an ecological and evolutionary perspective. Thus, it was never well-suited to be used as a framework for the evolutionary study of communication, meaning and language. But this evolutionary perspective has come more and more into focus during the last thirty years' attempts to bring together insights of neurosciences and evolutionary theory concerning the development of the human brain with cultural and social studies of the capability of early humans. These studies have been crucial for the integration of cognitive and brain sciences in the last ten to fifteen years. How did the success of the communicative abilities of early humans in their first primitive forms of language games influence the evolution of our brain? Terrence Deacon's book *The Symbolic Species* asks this question and tries to give answers on the basis of an interpretation in light of Peircean semiotics. He shows that the great survival value of the enlarged communicative capabilities as an instrument of social and cultural organization exerted a selective pressure on the development of the brain transforming the speaking ape into a (self)conscious human being.

This is some of the reasons why a new transdisciplinary and evolutionary semiotics has been focused on the evolution of consciousness and the production of communication and signification as well as on their interaction. These topics are still unsolved problems in the natural sciences and the humanities whose original framework was established before Darwin proposed his evolutionary theory and it became widespread in scientific explanation. We are therefore looking for a framework that can provide an ontology different from Cartesian dualism with its basic assumption of a mechanical nature separate from the sphere of the mind and of individual consciousness and different from the view that words and signs are passive instruments used by the human mind. It is well-known that C. S. Peirce, in his triadic, evolutionary and pragmaticist semiotic philosophy, offered an alternative that broke with mechanical determinism as well as with materialism. Through the first of his three categories, Firstness, he introduced *pure feeling* into the ontology of nature and *evolutionary love* as the driving force behind of the development of human life and consciousness. He also introduced a triadic concept of semiosis according to which, the Representamen is the expression of Firstness, the Object the expression of Secondness and the Interpretant the one of the mediating power of Thirdness. Through the inner life of Firstness, he gave the triadic signs a a dynamics of its own in the form of semiosis. Signs, in Peirce's theory of semiosis, are autonomous agents who have a life and an evolution of their own. Actually, Peirce saw the human self as a symbol in development through the accumulation of experience and the whole Universe in its cosmological evolution as an argument. Many researchers, among them Karl-Otto Apel and Jürgen Habermas (with a somewhat strange interpretation), have been attracted by Peirce's radically new way of thinking, and it has made a great impression on the Copenhagen School of Biosemiotics and its members Jesper Hoffmeyer, Claus Emmeche, Frederik Stjernfelt and Søren Brier. Nevertheless, it is obvious that Peirce's worldview is at odds with the standard view of nature, the nervous system and consciousness in the sciences as well as with the European tradition of phenomenology since Husserl. Thus, in our view, we are currently faced with the endeavour of reconciling Peirce's semiotics with mainstream views in which we find few pure Peircean semioticians. This is the approach from which we will present the invited articles to this special double issue. The topics of the articles concern fundamental and hitherto unresolved problems of Western knowledge, so when we give a critical evaluation of how far the various attempts have come in solving the problems and what remains to be done, it is not a personal criticism of the researcher for their failing, because no one has found a solution to these questions so far. The writers here are all established scholars and scientists. Some of them were invited to a workshop given at Copenhagen Business School in March 2008 some months after Søren Brier's inauguration as a professor in the semiotics of information, cognition and communication studies. Thus, we are giving our personal evaluation of the progress in this area that is so vital to the cybersemiotics Søren Brier has been developing.

    The major attempt at developing a new semiotics is then a transdisciplinary Peircean endeavour, as we see it in Winfried Nöth's article "On The Instrumentality and Semiotic Agency of Signs, Tools, and Intelligent Machines." The author underlines that Peirce's transdisciplinary semiotics breaks with a long tradition of instrumental theories of the sign, in which the sign used to be considered to be a passive instrument of thought and communicative action. Peirce's semiotics, by contrast, is based on the premise that signs, in the process of semiosis, are semiotic agents with a certain autonomy of their own. Nöth concludes that although instruments are signs, signs by themselves are not merely instruments but have an autonomy of their own in the Peircean perspective. This assumption differs from many of the received views of language philosophy, cognitive linguistics, computer semiotics, and evolutionary robotics. The article refrains from dealing with the further consequences of Peirce's evolutionary philosophy for the theory of consciousness and subjectivity.

Per Durst-Andersen, a Peirce inspired semiotically oriented cognitive linguist, contributes to the volume with the article "The Grammar of Linguistic Semiotics – Reading Peirce in a Modern Linguistic Light." He presents a new typology of linguistic signs based on Peirce's sign conception and his ten sign classes. Durst-Andersen argues that the symbolic nominal lexeme has an arbitrary relationship to its object, which makes it omnipotent in its openness to refer to anything within its own defined limits. But this omnipotence is at the same time its impotence. This is the reason why it must have a grammar as a vehicle that can give a semiotic direction to the otherwise completely static sign, which cannot by itself refer. He shows that the so-called grammemes are indexes and icons at the same time. He explains this by arguing that the inherent iconic and indexical properties of the symbol have been transformed into external functions by grammar. This is the only way in which the extrapolated symbol can be made concrete in discourse. Durst-Andersen also points to a third level where the verbal categories collaborate in order to make the dynamic arguments of deduction, abduction and induction corresponding to Peirce's argumentative signs. The theory thus gives us a completely new perspective on the relation between Peircean semiotics and linguistics. Still the theory is only a theory of linguistic semiotics, but an extension of it could be an important contribution to a full Peircean semiotics and useful to evolutionary theories of the development of the animal and human capacity for signification.

An important part of a new transdisciplinary Peircean semiotics is the development of a biosemiotics theorizing on the semiotic capabilities foundations in the biological makeup of all living systems (down to the single cell) as you can find it theorized in Kalevi Kull's article: "Biosemiotics: To Know, What Life Knows." He describes the field of semiotics as a general study of knowing and recommends following Rein Vihalemm's distinction between phi-sciences (that use physicalist methodology) and sigma-sciences (that use semiotic methodology). Kull analyzes Jesper Hoffmeyer's book *Biosemiotics* as an example of inquiry into the sigma-scientific approach to living systems. Also, Kull struggles with how to construct an evolutionary theory that can include the first person position as a real agent. He describes the complementarity of Jakob von Uexküll's theory of meaning with a description of the evolution of life from a physical chemical background. Where Kull seems to avoid Peirce's metaphysical synechism Jesper Hoffmeyer has struggled with ways of adapting it to a general semiotic biology, partly using Gregory Bateson's work in cybernetics to play down Peirce's metaphysical categories especially the role of pure feeling in Firstness. However, a theory of first person consciousness and its role in the production of meaning still remains a central problem in the metaphysical founding of biosemiotics, especially since Marcello Barbieri has promoted his code semiotic theory that prefers to see biosemiotics as an informational extension of the standard biological and evolutionary paradigm only later to introduce Peircean triadic semiotic in animals with a consciousness based on a central nervous system. Thus Barbieri just takes consciousness as a given fact, but avoids any deep theory of how it emerged through evolution. Instead he gives agency to code-making molecules for instance in the cells genetic machinery.

The next researcher in this volume, namely floyd merrell, boldly investigates the full consequences of Peirce's theory of musing and Firstness in comparing it to one of the most famous and influential Buddhist philosophers of all times, namely Nargajuna of the second century. The title of merrell's metaphysical investigation is: "Musement, Play, Creativity: Nature's Way." The author underlines that Peirce's notion of musement, from which he developed his theory of abduction and signification, is based on a nonlinear, nonbivalent, plurimorphic process view of the deepest level of reality. Knowing how interested Peirce was in Buddhist philosophy, it may not be so surprising as it may seem, that merrell finds part of Peirce's philosophy expressed and further developed in the Buddhist philosopher Nāgārjuna's founding philosophy of the Middle Way. In his analysis, merrell shows how Peirce's concept of signification is based on the importance of play and creativity in basic reality and therefore also is the natural consequence of musement. Semiosis is no simple deterministic process, neither in Peirce's nor in Nargajuna's philosophy of emptiness. It is interesting to know that emptiness and triadicity are basic concepts also in George Spencer-Brown's cognitive philosophy of making a distinction in his seminal work: *Laws of Form*.

Another major attempt to formulate a transdisciplinary semiotics under development goes under the heading of cognitive semiotics—with a journal by the same name. In his contribution to this volume Göran Sonesson's proposes a new development of phenomenological theory combining ideas from Edmund Husserl and Aron Gurwitch with elements from Piaget's theory, and attempts to insert this framework into an evolutionary frame. On the level of human communication then Peirce's triadic semiotics concepts such as Icon/iconicity, Index/indexicality and Symbol/symbolization are used in more differentiated ways than Peirce did, but in an attempt to avoid founding the paradigm on Peirce's objective idealist and synechistic ontology and his three basic phaneroscopic categories. Both Sonesson and Jordan Zlatev in his own contribution find that it is too metaphysically demanding to take over Peirce or any other theoretician wholesale and prefer developing their theories step by step from data analysis. But—one is tempted to say—at the cost of basing themselves on a common sense realistic view within the received view of science. Göran Sonnesson's article: "The View from Husserl's Lectern. Considerations on the Role of Phenomenology in Cognitive Semiotics" represents a nearly mature stage in developing such a framework including an advanced version of cognitive science as part of a large program in cognitive semiotics at Lund University. Sonesson attempts to avoid common sense by taking a phenomenological point of departure. One of the major obstacles in such a work is of course to make a standard materialistic view of evolution based on thermodynamics and Darwinian biology produce a theory of first person consciousness that is the basis of the phenomenological aspect of the theory. Sonesson tries to avoid this problem through a phenomenological view on the contribution of science, but quite how this is to be done is only sketchily suggested in the article, as integrating evolutionary theory into his framework is a relative late aspect in its development. Finally it is also a framework problem to make intersubjective consciousness a basic part of the paradigm following up on the late development of Husserl's paradigm on this matter.

In his article in the present volume: "Levels of Meaning, Embodiments and Communication" Jordan Zlatev—from a slightly other version of cognitive semiotics than Sonesson's—presents a detailed analysis of various types or rather conceptions of embodiment and communication. This is done within an evolutionary framework in combination with a classification of semiotic stages into a semiotic hierarchy. He works with a biological, a phenomenological, significational (sign-based and an extended/normative) level. Zlatev operates with theories based on first, second and third person's observation, but the theory still has not yet reached an explanatory stage concerning the emergence and ontology of consciousness in an essentially physicalistic framework recognizing first person experience as being as real as matter; the author has so far refrained from declaring such a philosophical ambition. Still, his work adds important aspects to the greater picture.

Ranulph Glanville's column *A (Cybernetic) Musing: Design and Cybernetics* takes it s point of departure in Gordon Pask's article "The Architectural Relevance of Cybernetics" and his *conversation theory*. The critical connection between cybernetics and design in the context of design is understood as a reflexive conversation with the situation. The column develops the consequences of this view for architecture and design.

The ASC column for this issue is provided by one of the Trustees of the society, Paul Pangaro, and two colleagues, Hugh Dubberly and CJ Maupin. The paper entails an application of basic cybernetics concepts to a perspective on behavior; namely it deals with the notion of bio-cost as the economics of human behavior. Bio-cost is explained in terms of energy, attention and stress as a function of time, distinctions that are normally difficult to quantify and thus to effectively include in business or design conversations. The authors present circular views of the bio-cost and the resultant gain experienced by single and multiple participants in achieving their goals. They claim that paying attention, and choosing to reduce bio-cost, can result in greater efficiency, capacity and variety in human behavior.

The two books reviewed in this issue are both collections of essays by a number of contributors, and both books address the same overriding topic—the implications of the concepts of information and computation to our scientific and philosophical understandings of the world. The first book, *Computation, Information, Cognition: The Nexus and the Liminal*, was edited by Susan Stuart and Gordana Dodig-Crnkovic. It is reviewed for us by Vincent Müller. The essays in this book were presented in an earlier form at the European Computing and Philosophy Conference held in Sweden in 2005. The second book, *Philosophy of Computing and Information—5 Questions*, is here reviewed for us by the same Gordana Dodig-Crnkovic who co-edited the first book. In this latter book, the editor, Luciano Floridi, poses a series of challenging and open-ended questions to a series of eminent thinkers, of whom the most famous are probably Dennett, Dreyfus, Searle, Winograd and Wolfram. Both books taken together demonstrate how much the concepts of information and computation have been integrated into cutting edge science from physics to biology to linguistics. This has encouraged a renewed interest in these concepts by philosophers as well. Yet philosophical analysis ends up illustrating what is perhaps obvious even to a non-philosopher—that these concepts do not themselves remain philosophically stable as they are applied by different researchers in different fields of study.

The artist for this issue is Doug Craft. He is inspired by the organization of form that he perceives in nature, and is particularly drawn to fractional symmetries, self similarity and the Golden Ratio. The proportioned branching of the latter can be seen in trees, feathers, and microscopic crystals. Several of the images in this issue are microphotographs of crystals. The cover mandala is a formalization of overlapping Golden Rectangles which photographically represent the classical basic elements: fire, earth, air and water. See http://tinyurl.com/C-HK-artists for further explication.

Craft, D. (2009). *Cultural Clash*. Microphoto: Benzoic Acid Melt Crystals; 34 x 21 cm.

Cybernetics and Human Knowing. Vol. 16, nos. 3-4, pp. 11-36

On the Instrumentality and Semiotic Agency of Signs, Tools, and Intelligent Machines

Winfried Nöth[1]

In instrumental theories of the sign, from antiquity (*Cratylus*) and the Middle Ages (Scholasticism) to 20th century semiotics, signs have been defined as tools of thought and communicative action. C. S. Peirce's semiotics is based on the anti-instrumental premise that signs, in the process of semiosis, are not instruments but semiotic agents acting with a semiotic autonomy of their own. Although instruments are signs, signs are not mere instruments. The "user" of a sign cannot act with it as an instrument because the sign has an autonomy of its own, which the instrument has not. The relevance of this radical theory of semiotic agency to the semiotics of computers and robots is examined in the current contexts of language philosophy, cognitive linguistics, R. Millikan's teleosemiotics, computer semiotics, and evolutionary robotics.
Key words: signs, instruments, machines, C. S. Peirce, R. Millikan, semiotic agency, semiotic autonomy, computer semiotics

1. Signs as Instruments of Thought and Action

That signs are instruments of those who use them for various purposes sounds like a truism. Not only do we hear it in everyday metaphors of verbal communication, but the instrumental view of the sign is also widespread in the history of semiotics since antiquity. We find it in the philosophy of language, in communication theory, and in media studies. There are few who argue against it, but when it comes to determine what is used by whom and for which purpose, the answers are diverse and in some respects even controversial. The argument that signs are instruments of communication has few opponents (but see section 2.). Less accepted is the view that signs are also instruments of thought and cognition.

1.1 Signs as Instruments of Communicative Action
The argument that verbal signs are used as instruments has been defended since antiquity. It is central to one of the earliest treatises on the philosophy of language, Plato's *Cratylus*. In this dialogue on the relation between words and the objects which they stand for, Socrates compares the practical utility of the craftsman's tools with the utility which words have for a speaker. A tool, says Socrates, is naturally adequate and appropriate to the task for which it serves. To the weaver, the "right and natural" instrument (Plato, 1953, 391e) is the shuttle, to the blacksmith, the anvil and the

---

1. Universität Kassel, FB 02, D 34109 Kassel, Germany. Email: noeth@uni-kassel.de
   Website: http://www.uni-kassel.de/~noeth

hammer serve best to shape the iron, and to the shoemaker, the awl is the natural tool for piercing leather. As any practical action requires its proper tool, so does the act of speaking. The speakers' appropriate instruments for communicating their ideas are their words (Plato, 1953, 388a-389c). Practical instruments have a practical utility, whereas words serve a semiotic utility. As the weaver's shuttle is useful for separating the warp from the woof, so is the word useful to the speaker, who wants to communicate or to organize his or her ideas.

In the history of semiotics, this early instrumental theory of the sign has lived on in many variants. The Scholastics and Late Scholastics had a much discussed theory of semiotic instrumentality according to which all signs by which we communicate acoustically or visually are instrumental signs. After the Scholastics, the theory of the instrumentality of signs in communication fell largely into oblivion. In the age of rationalism, semiotics was more interested in the study of signs as representations of things and ideas than in communication; and to the mentalist semioticians of 20th century structuralism, signs were elements of a system and not really instruments to communicate ideas.

The linguistic turn in 20th century philosophy brought the instrumental view of language back on the agenda of language philosophy (cf. Pagee, 1967). Wittgenstein (1953) revives the Socratic topic in his *Philosophical Investigations*. *Instrument* and *tool* are appropriate characterizations of verbal signs in his philosophy of language since it postulates that the proper study of verbal signs is the study of their *use*. What language and instruments have in common is that they are both *used* and have *functions*: "Think of the tools in a box: there is the hammer, pliers, a saw, screwdriver, a glue-pot, nails and screws.—The functions of words are as diverse as the functions of these objects" (Wittgenstein, §11). Whereas Socrates focuses on the utility of verbal signs in communication, Wittgenstein's focus is on their meaning and the way meaning is revealed in the use of signs. Not only language in general is an instrument, but also "its concepts are instruments" (§569). Instruments as well as language have their sense in their employment: "Look at a sentence as an instrument and its sense as its employment" (§421).

In 20th century semiotics, the functional views of language and communication were based on instrumental premises. According to the tenets of Prague structuralism, signs serve social and semiotic purposes. As *instruments of communication*, verbal and nonverbal signs are used to fulfill the communicative needs of the sign users, whose purpose it is to transmit a message and thus to establish and maintain social contacts. It is the semiotic function of a sign to be relevant, that is, to be distinguishable from all other signs of the system in order to make the interpretation of the message possible. Signs are instruments because they serve the speakers' communicative purpose, and these instruments are adequate to the degree that they are relevant (cf. Nöth, 2000, pp. 201-202).

In Vienna, Karl Bühler proposed an explicitly instrumental theory of the sign. With direct reference to Plato's *Cratylus*, he made the instrumental theory of the sign a cornerstone of his *organon model* of verbal communication. Language, according to

Bühler (1934, p. 24), is a tool (Gr. *órganon*), which speakers use to influence their addressees cognitively or emotionally.

In Geneva, a formalist instrumental theory of the sign was proposed by the semiotician Luis Prieto in the 1960s and 1970s. According to Prieto, the utility of instruments and practical tools consists of the class of operations which can be performed by means of them (Prieto, 1966, pp. 7-8). Whereas practical tools serve the purpose of operating with effects on the exterior world, signs are instruments of the human mind. Their purpose is to be useful in conceiving, understanding, and recognizing the exterior world and to serve communicative needs. Prieto identifies the word in its instrumental function with the Saussurean *signifier*, that is, the verbal sign in its spoken or written manifestation, and the purpose of this instrument with its *signified* (Prieto, 1966, p. 9), redefining the latter as "the class of influences which the speaker can exert [on the hearer] by means of" the former (Prieto, 1973, p. 153). Furthermore, he distinguishes between the instrumentality of natural and artificial signs: although both serve as tools for a purpose, natural signs, such as clouds which stand for rain, and natural tools, such as sticks or stones, are not produced for the purpose which they serve; only artificial signs are instruments produced for a semiotic purpose.

Prieto's instrumental theory of the sign was one of the few semiotic theories from the West that could be discussed and studied in East Germany, where it appeared in translation (Prieto, 1972). Apparently, the instrumental theory of the sign seemed compatible with Karl Marx's theory of how tools function in the transformation of nature through human labor. In fact, Marxist semiotics developed its own instrumental theory of the sign along these lines. Vygotsky and Rossi-Landi are two representatives in this tradition.

Vygotsky (1981) created the term *psychological tool* to characterize spoken or written words, numbers, diagrams, maps, and other conventional signs as mediators between external stimuli and internal cognitive responses. According to Vygotsky, signs are mediators in our otherwise unmediated and direct behavioral interaction with environmental stimuli, which serve as "instruments directed toward the mastery or control of behavioral processes" (Vygotsky, p. 137). Signs are the "tools of an instrumental act" (p. 139) in which they serve as "a means with the help of which we direct and implement the psychological operations (memory, comparison, selection, etc.) necessary for the solution of a problem" (p. 139). Sign users, according to this model, are autonomous semiotic agents who make use of signs as instruments for nonsemiotic (behavioral) purposes.

The distinction between the sign as an instrument and its nonsemiotic purposes is also essential to Rossi-Landi's instrumental theory of the sign. The Italian Marxist semiotician argues that verbal signs are like practical tools since they serve purposes "whose ends lie outside of them" (Rossi-Landi, 1974, p. 1873). The homology between sentences in language and material tools and the parallelism of their evolution in human culture is due to their indissoluble linking in the course of production and labor. As practical utensils possess in themselves "the form of the use or of the working process to which they lend themselves" so do sentences as verbal utensils (Rossi-Landi, p. 1872).

Despite the different backgrounds against which signs have been defined as instruments of communication, there are common denominators. Firstly, instrumental theories are opposed to representative theories of the sign. Both distinguish between the *aliquid* and the *aliquo* of the sign relation, but whereas the representative theory of the sign defines the nature of this relation as a *standing for*, the instrumental theory considers this relationship as one of *serving the purpose of*. Secondly, instrumental theories of communication tend to establish a dualism between the semiotic instruments of communication and biological, psychological, or social effects for whose nonsemiotic purpose they are used. The metaphor of the instrument supports this dualist way of interpreting the effects of signs. Tools have a finality which is not tool-like. As tools are means for purposes which are not tools, so are signs considered to be a means for nonsemiotic purposes in the framework of instrumental theories of the sign. This dualism with which most instrumental theories distinguish between signs and the nonsemiotic purposes for which they are used is in sharp contrast with Peirce's anti-instrumental theory of semiosis (see section 2.).

1.2 Are Mental Signs Instruments of the Mind?
The view that signs are instruments not only for the purpose of influencing others but also of private thinking is discussed in Plato's *Cratylus*. Socrates expounds it when he says that signs are useful instruments for two other purposes besides the one of communication, namely, giving names to objects and "distinguishing things according to their natures" (Plato, 1953, 388b). Translated into the terminology of modern semiotics, Socrates addresses with these arguments the semiotic functions of representation and classification (cf. Keller, 1995, p. 29). According to this view, mentally used language is a useful instrument for cognition and orientation in life.

Without reference to Plato, the Scholastic semioticians argued against the view that mental signs are instruments of cognition. In their semiotics, *signa instrumentalia* were opposed to a class of noninstrumental signs which they called *signa formalia* (Meier-Oeser, 1997, pp. 235-254; Deely, 2001, pp. 893 & 913). The latter are the ones which Peirce, centuries later, called *thought-signs*, that is, mental representations which constitute our thought and which we use in our internal dialogues with ourselves. Whereas instrumental signs serve to communicate our ideas to others, formal signs are concepts of the human mind. The medievals called them *formal* because they *form* our cognitions and thus *give form* to our ideas (cf. Nöth, 2000, p. 12). Implicit in the medieval distinction between instrumental and formal signs is the assumption that ideas or thought-signs do not serve as instruments of thought. Who should be the agent using these signs, and for which purpose? The only purpose which thoughts serve is the purpose of thinking. Since thinking is nothing but the use of thought, it would be either tautological to say that thoughts are instruments of thought or it would lead to an infinite regress involved in the idea that thoughts are the instruments of thinkers: if all thoughts are the product of a thinker but thinking is constituted by thought, where do the ideas of the thinker come from? Thoughts, in sum, constitute both thinking and the thinker. They cannot be said to be semiotic instruments.

1.3 Verbal Signs as Instruments of Thought

While Plato and the Scholastics discussed the question whether thoughts are instruments of thinking, a different question concerning the relation between thoughts and signs is being debated in modern cognitive psychology. It is the questions whether *public* signs are instruments of *private* thought. In this context, the term *instrument* is sometimes only implicitly used with reference to the verbal expressions of ideas. The question is whether thoughts precede their articulation *by means of* speech or writing. Logan (2007, p. 90-113) gives a survey of the diverse positions in this controversial debate. Jerry Fodor (1975) is the most prominent scholar among those who argue that words are instruments by means of which we express ideas, which we have before they are verbally expressed. In other words, the *language* of thought is not the language in which we express our thoughts. Fodor postulates a universal language of thought, which he calls *mentalese* and which differs from the language of public speech, by means of which it is expressed in an individual language, such as English, French, or Chinese. Other authors who adopt similar positions are Pinker (1999) and Donald (2001). According to the latter, public "symbols get their meaning from thought" and are "invented in the service of thought" because "humans can think without language" (Donaldson, p. 276-277).

Among those who contest this view of the instrumentality of public language, defending the counterargument of the unity of thought and language is Carruthers's (1996, p. 57), who argues that "it can plausibly be denied that there is any determinate thought in existence prior to its linguistic formulation." Words are hence not instruments to express thoughts or ideas, because one does not *first* entertain a thought and *then* express it by means of words. Instead, thinking *is* the mental processing of the same signs which we use in public communication. Peirce formulated these insights a century before it began to be discussed in current cognitive psychology (see below, section 2.).

1.4 Metaphors of the Instrumentality of Signs Mediating Signs

The view that signs are instruments for the purpose of transmitting, disseminating, translating, or circulating signs can be found in everyday metaphors of communication, in communication theory, and in media studies.

If words are instruments of communication, money is the instrument par excellence of economical exchange, and money as a metaphor of language is a metaphorical concept expressing the idea of the instrumentality of language. It was Leibniz who called the word a reckoning penny (*Rechenpfennig*), but the metaphor is not restricted to philosophical discourse. The metaphor that words are coins (Weinrich, 1958) is rather common in everyday language: Language is money, and communication is the more or less profitable exchange of ideas.

Another everyday view of the instrumentality of communication is apparent in metaphors which represent words as objects which must be transported from a sender to a receiver. Reddy (1993) described it as the conduit metaphor: The message must be put into words, stored in a container by a sender, transmitted by a messenger via channels to a receiver, and extracted from the container by its receiver. In this scenario, the message is the sender's instrument, and the medium conveying the message, a letter, a newspaper, the radio, or more generally speech, writing, or pictures, is a kind of meta-instrument, an instrument to deliver the instrumental sign.

Classical communication theory basically adopted the conduit metaphor in its sender-message-receiver models of communication. The logic of the communicative scenario underlying this model presupposes that messages require messengers to *convey* them and that such a messenger is the instrument of its sender. According to the well known Shannon/Weaver model, the signs of the message are selected from a sign repertoire (or code) by a sender; these signs are then translated into signals, which, in turn, are transmitted via a channel to the receiver. The metaphorical messengers employed in this scenario of communication are the signals and the medium. Information theory distinguishes between the *message* and the *signal* transporting the message via a channel from a sender to a receiver; its premise is that the signal is the mere instrument for transmitting the message and that only the message is the sign (or *symbol*). Hence, signals serve to transport and to deliver signs. However, the logic of signs requires that the signals encoding a message must be signs, too, for how else could they be decoded if they had not meaning to be decoded? In other words, signals are signs conveying signs, that is, instrumental signs.

1.5 The Instrumentality of the Media and of Writing
It seems to be a truism to say that the media are instruments for the dissemination of ideas. Often, the expression *using the media* has negative connotations. Propagators of biased or self-interested messages are criticized for misusing or abusing (G *instrumentalisieren*) the media when using them in illegitimate ways as tools for the propagation of their programs. Whether the media are indeed instruments or not, such ways of referring to them testify to an instrumental view of their message, which are seen as tools serving their senders' interests to the advantage or disadvantage of the masses influenced by them.

Marshall McLuhan has argued against the idea of mere instrumentality of the media in his much quoted dictum that "the medium is the message." The *content* of a medium, as McLuhan (1964, p. 8) puts it, "is always another medium" since each new medium invented in the course of the history of the media made explicit or implicit reference to its precursors, radio to newspaper, television to radio etc. If a medium as such has content, as McLuhan states, it must be more than a mere instrument of message transportation; it must itself be a sign, and the message which the medium conveys as a medium is not the message of the one who uses the medium. In addition to conveying a message about stories and events, the medium also conveys a self-referential message about itself, and in this respect the media are not instruments (cf. Nöth, 2007).

A medium whose instrumentality has often been claimed is writing. English proverbs, such as "Pen and ink is the wit's plough" or "The pen is the tongue of the hand," ascribe to the medium of writing (via the metonymy of the hand and the pen as its instrument) the status of an instrument of the human mind. The instrumental view of writing is also inherent in the view that writing is a secondary code serving as an instrument to represent speech (cf. Nöth, 2000, p. 359). Arguments against such *phonocentric* views of writing (as Derrida has called them), which interpret written words as instruments of thoughts whose substance is silent speech, have been raised in media theory, the semiotics of writing, and in cognitive psychology. According to McLuhan's media theory, writing is a medium which has its own *content*; written texts are not only the expression or translation of spoken ones. In the semiotics of writing, it has been shown that ideographic writing systems do not really represent speech sounds at all, but are autonomous sign systems for the direct representation of ideas. Even alphabetic writing systems have become largely independent from the spoken form of the language for whose representation they were originally invented.

The cognitive relationships between speaking and writing have equally been interpreted in terms of instrumentality. A rather common view of this relation is the following: We write down thoughts which precede their expression in writing; these thoughts are expressed in the form of a silently spoken language. Writing is thus conceived of as the instrument of conveying the signs of a silently spoken language to an addressee. Carruthers (1996, p. 50), for example, describes the process of writing a letter as follows: "When I sit and draft a letter to someone in my head, for example, what figures in my consciousness is the sequence of English sentences in auditory (and perhaps kinaesthetic) imagination, rather as if I were dictating that letter aloud." Nevertheless, cognitive psychologists have also given evidence of a certain autonomy of writing in relation to speaking. Carruthers, despite his account of writing as a representation of inner speech goes on to give arguments which also underline the cognitive independence of writing from speaking. Professional writers do not first think and then write: there is a thinking on paper or at the keyboard during which "one does not *first* entertain a private thought and *then* write it down; rather, the thinking *is* the writing" says Carruthers (1996, p. 52; cf. this paper section 1.3). Writing thus described is not an instrument which expresses spoken words; the written word is the expression of a thought whose mental form is the image of a sequence of letters (see section 2.1).

1.6 Symbols as Existential Tools of the Human Mind
From the rather different perspectives of phenomenology and anthropology, the media philosopher Vilém Flusser (1998, p. 76) defines human symbols as instruments. According to Flusser (1996, p. 76), we communicate in order to overcome the abyss which separates human culture from nature. In the resulting fundamental situation of human *ex-istence* in the literal (etymological) sense of standing outside (of nature),

symbols are artificial instruments, mediators, and bridges over the gulf between culture and nature.

Instruments are extensions of the human body, as Flusser (1997, p. 22) observes: "The arrow extends the finger, the hammer extends the human fist, and the hoe extends the toe." In such extensions of the human body by means of tools, it is not the tool that has the role of an agent but the tool user. In the process of writing, pen and paper are evidently the writer's instruments in this sense, but the words which a writer writes by means of them are instruments too, although of a different kind. It was Peirce (see section 2.) who described paper and pen as necessary but not sufficient instruments (acting as merely efficient causes) in the semiosis. Whereas the sign involves the triadic interaction of an object, a sign, and its interpretant, instruments merely involve the dyadic interaction between the instrument extending the agency of its user and the practical effects achieved by means of it. The instrument is a merely efficient cause of this effect, which could be reached by other means. A pen is an instrument which serves its writer as an efficient cause to produce a sign since the writer can use other pens to write the same message.

2. Peirce's Arguments Against the Instrumental Theory Of Signs

Peirce's semiotics is basically incompatible with the view that signs are instruments. Neither thought signs nor signs used in communication are instruments of those who think or communicate; signs are neither instruments of thought nor are they genuine instruments of communicative action. Above all, publicly expressed signs are not the instruments by which private thought signs are expressed, for "thought and expression are really one" (CP 1.349, 1903). The sign is inseparable from its object, but unlike the Saussurean signifier, which is the phonetic expression of a mental concept (see section 2.3), the sign according to Peirce is not an outward expression of an inner mental concept or thought-sign. Both ideas (as private thought-signs) and outward public signs are signs which have their own objects and interpretants. The semiotic unity of sign and object precludes, for example, that good ideas may suffer from being expressed in bad language, for "it is wrong to say that a good language is *important* to good thought, merely; for it is the essence of it" (CP 2.220, 1903). The sign in good language differs from the sign in bad language; a sign can only represent its own object, not the object of another sign.

Peirce's explicit and implicit anti-instrumentalist positions may surprise at first sight because his semiotics is based on two key terms which sound compatible with the instrumentalist theory of the sign, *medium* and *purpose*. However, a closer examination of these concepts reveals that neither of them has any instrumental connotations. Nevertheless, although the sign is not an instrument, instruments are necessary for the propagation of signs in the sense that we need a pen and paper to write down words. The conclusion that signs are essentially not semiotic instruments does not preclude the reverse: instruments, for their part, are in fact signs.

2.1 The Sign as a Mediator, Semiotic Causality, and the Instruments of Semiosis
Peirce defines the sign as a *medium* (MS 339, 1906), but not as one which mediates between an addresser and an addressee. Instead, the sign mediates between its object, which it represents, and its interpretant, which is its interpretative effect. The agent in the process of semiosis in which the sign creates an interpretant, is the sign, not the addresser, and the agency of the sign is one of final causality: it is the *purpose* of the sign to create an interpretant. In the process of semiosis, the sign is determined by its object, since it must represent it to be its sign, and it determines its interpretant, since the latter is created by the former.

Peirce distinguishes between efficient and final causality (Pape, 1993; Santaella, 1999). Instruments contribute to the process of work in which they are uses as efficient causes. A hammer and an axe are instruments whose effects involve efficient causation. Efficient causality describes the dyadic cause-effect relation by which a cause brings about its effect *blindly* or, as Peirce put it, by "brute force," in a process determined by irreversible physical laws (cf. CP 6.329, ca. 1909). The force of efficient causality is "a compulsion determined by the particular condition of things … acting to make that situation begin to change in a perfectly determinate way" (CP 1.212, 1903). Efficient causality and hence also instrumentality belong to the category of secondness, the category of dyadic relations which leave no alternatives for creative change, whereas signs belong to the category of thirdness.

Signs, by contrast, operate by final causality, even though they cannot do without efficient causes to convey their messages. Final causality involves triadic interaction; it is the long term causality of purposes, intentions, ideas, signs, and general laws, all of which belong to the Peircean category of thirdness. Final causality "does not determine in what particular way [the general result] is to be brought about, but only that the result shall have a certain general character" (CP 1.211, 1902). Signs act by final semiotic causation insofar as their semiotic potential can reach its effect by different means. For example, ideas (which are signs) spread in the memory of different brains, propaganda may be propagated by many copies of the same poster; the same message may be carried by different messengers.

Each final cause needs an efficient cause to reach its effect. In semiosis, this principle explains the potential of semiotic creativity. The same sign may be propagated by means of different semiotic instruments. Among the efficient causes of verbal semiosis are the sound waves of vocal communication and the printing press which multiplies the message of a journalist (CP 1.213, 1902). The thought or action resulting from the interpretation of a sign is the interpretant of the sign. To create an interpretant is the sign's final causality. A sign cannot reach this purpose without its phonetic or graphic materiality, which is necessary in any process of communication to convey the sign's message. This materiality is the efficient cause of the sign. Not only the material medium but also the material tools of writing, such as the inkstand used by a writer to write down an author's idea, and even the neurons of our brain, which make thought possible, are examples of efficient causes in the process of semiosis given by Peirce (CP 7.366, 1902).

In sum, signs are not semiotic instruments themselves, but semiotic instruments, such as our voice, tongue and lips, pen, paper, typewriter, telephone or computer are necessary instruments and efficient causes in sign processes. In this sense, semiotic instruments are not only necessary for the purpose of communicating; thought-signs equally require semiotic instruments, and their instruments are our neurons, brain lobes, and our perceptual organs, our eyes and ears. This is probably Peirce's most radical argument: the brain is not the cause of the signs produced by speakers or thinkers; it is the instrument of the signs. Signs use brains to convey their messages. Living signs in the process of semiosis use the neurons of their sign users as their instruments and efficient causes for the purpose of their procreation.

2.2 Purpose, Semiotic Autonomy, and Circular Causalities in Semiotic Co-evolution

Peirce's concept of *purpose* differs from the one used by the instrumentalists. Whereas, according to the instrumental theories of the sign, purpose is the goal of a sign user to achieve a certain effect in an addressee, in Peirce's theory of semiosis, purpose is a goal which the sign has itself. The sign has a purpose of its own, which is not the intention of an autonomous sign user; it is the purpose to create an interpretant, that is, to "be interpreted in another sign" (MS 1476, 1904). Public signs have the purpose of creating interpretants in interpreters. The purpose of private thought-signs is to act in the mental dialogue in which a sign is "translated or interpreted in a subsequent one" (CP 5.284, 1868); it is hence the purpose of creating thought-signs, which are their interpretants. The purpose of public signs is to create interpretants by spreading among sign users. In contrast to instruments, which serve noninstrumental purposes, the purpose of a sign is hence a semiotic one.

Having its own purpose, which is the purpose of representing its object and of producing an interpretant, the sign is an autonomous semiotic agent acting with an autonomy absent in mere instruments. Evidence of the autonomy of the sign in relation to the sign user is first of all that it is determined by its object (and not by its user), for the sign does not only represent its object to the mind of an interpreter; the object represented by the sign is also "in a sense the cause, or determinant, of the sign even if the sign represents its object falsely" (CP 6.347, 1909; cf. Parmentier, 1985). To justify the autonomy of the sign by its determination sounds paradoxical, but what is meant is a kind of self-determination, since the object is part of the sign, and there is no sign without an object. The sign's self-determination thus defined means also that the purposes of the signs do not depend on the purposes of its individual users.

The way in which a sign is determined by its object depends on its modes of relation to its object and its interpretant. Rhematic signs, for example, are determined to represent their objects as such, without affirming, questioning, or negating their existence. Dicentic signs which affirm are determined by the truth of what they represent, and arguments are determined by the laws of logic and reasoning in general. Symbols are determined by the habits of their users, by which they have become signs. Indices are determined by their objects because they are related to them by causal, temporal, or spatial relations, whereas icons represent their objects by means of qualities of their own. All these determinants determine the sign independently of their users. For example, a rhematic symbolic legisign such as the word *apple* uttered by a speaker is determined by the habits of the speakers of the English language to represent the fruit of the apple tree; the word represents its object as a general kind and without affirming its existence, irrespective of whether the utterer has the intention to produce this representation or not. Of course, the sign user may use the word to refer to a pear at some particular moment, but in the long run, words cannot be used for purposes incompatible with the one they have independently of those who use them. A sign user may also lie or argue against the laws of logic, but to ignore the truth and the laws of reasoning will, in the long run, not pay. Other respects in which utterances are determined independently of their utterer are by the rules of grammar and the scope of the lexicon which restrict the range of syntactic and lexical choices. Furthermore, there are the biolinguistic and evolutionary constraints which have determined and restricted the possibilities of articulation and perception (cf. Deacon, 1997, p. 116).

To the degree that the sign is determined by its object, it is so to speak itself *used* by its object for the purpose of its representation, and the so-called sign users are used by their signs to convey the message of their signs. Insofar as sign users are determined by the logic of their signs and the nature of the objects which they represent, they cannot claim that they use signs as their instruments; instead, they are themselves being used by the signs which they use since they are the instruments of the sign's mediation. The circularity involved in this semiotic interaction between sign users using signs and signs using sign users testifies to a co-evolution of signs and those who produce them, which Peirce describes as follows:

> Man makes the word, and the word means nothing which the man has not made it mean, and that only to some man. But since man can think only by means of words or other external symbols, these might turn round and say: "You mean nothing which we have not taught you, and then only so far as you address some word as the interpretant of your thought." In fact, therefore, men and words reciprocally educate each other. (CP 5.313, 1868)

If signs are autonomous semiotic agents with purposes of their own, what is then the role of the sign users? Do they have no purposes and no intentions? Peirce solves the apparent paradox of the conflict between the purposes of the sign users and those of the signs with the argument that sign users are themselves signs so that their intentions are, in turn, purposes of signs. By no means do sign users pursue nonsemiotic purposes, for they themselves are and act as signs in the flow of thought-signs and public signs which constitute them as signs. This is the line of argument of the famous passage in which Peirce defines the human being as a sign:

> Without fatiguing the reader by stretching this parallelism too far, it is sufficient to say that there is no element whatever of man's consciousness which has not something corresponding to it in the word; and the reason is obvious. It is that the word or sign which man uses is the man himself. For, as the fact that every thought is a sign, taken in conjunction with the fact that life is a train of thought, proves that man is a sign; so, that every thought is an external sign, proves that man is an external sign. That is to say, the man and the external sign are identical, in the same sense in which

the words homo and man are identical. Thus my language is the sum total of myself; for the man is the thought. (CP 2.314, 1868)

In sum, the sign user's intentionality is not the sole cause of a sign process. In so far as it is determined by its object, a sign is not the intentional product of an autonomous sign user. The addressee is determined as much by the sign's purpose as by the sender's intention, but the latter is the purpose of a sign itself. Signs, and in particular words, "only stand for the objects they do, and signify the qualities they do, because they will determine, in the mind of the auditor, corresponding signs" (CP 2.92, 1902).

2.3 Brief Excursus on Saussure's View of the Autonomy of the Sign

Despite the fundamental differences between Charles S. Peirce's semiotics and Ferdinand de Saussure's semiology, the topics of the autonomy of the sign and its anti-instrumentalist interpretation are issues of common concern to both theoreticians of the sign, although they develop their approaches from rather different perspectives. The value of a sign, according to Saussure (1916/1959, p. 114-115), is solely determined by its relation to other signs of the sign system to which it belongs. The sign is not an instrument for anything else, since "without language, thought is a vague, uncharted nebula" (Saussure, p. 112). Signs are thus autonomous semiotic structures determined by no other purpose than the one of making a semiotic difference which allows the sign to be distinct from other signs expressing other concepts. Above all, the signifier of a sign is not the instrument expressing the signified of the sign let alone its object of reference (*pace* Rossi-Landi, see 1.1), for the signifier and the signified are mutually determined like the recto and the verso of a sheet of paper: "Thought is the front and the sound the back; one cannot cut the front without cutting the back at the same time" (Saussure, p. 111).

2.4 Excursus on Millikan's Teleosemiotics

In the framework of modern cognitive science, quite independently of Peirce and largely without reference to his pioneering ideas on the topic, an evolutionary theory of the sign has found wide resonance which seems to have remarkable affinities with the teleological aspect of Peirce's theory of semiosis. Kappner (2004, pp. 279-311) gives a survey of the main ideas of the sign theory of this topical trend in cognitive science, which Millikan calls *teleosemantics*, Dretske *teleofunctionalism*, and Papineau the *teleological theory of representation*. Kappner also discusses some affinities and differences between these trends in cognitive science and Peirce's semiotics in light of Short (2007). The main representatives of this teleological theory of signs and language on evolutionary principles are Dretske (1983, 1988), Papineau (1987), Neander (1995), and Millikan (1984, 2004, 2005; see also Rolf, 2008). The following excursus is restricted to the influential writings of Charles Morris's former student Ruth Millikan. The general theory of signs based on Darwinist principles of evolution which Millikan calls *teleosemantics* is actually a *teleosemiotics* since it is by no means restricted to the study of verbal signs but also deals with all kinds of meaningful cultural objects including tools and instruments (cf. Millikan, 1984, p. v).

In contrast to the mainstream doctrines of intentionality in verbal behavior defended by Chomsky, Fodor, Searle, or Grice, which define speakers' intentions as the sole cause of language utterances, Millikan, like Peirce, ascribes autonomous purposes to the signs by which speakers communicate. The purpose which a sign has, according to Millikan, is its *proper purpose* or *proper function*.

Millikan's teleosemiotics is rooted in the Darwinist theory of biological evolution. She argues that ultimately all signs, even intentional verbal signs, have evolved from the natural signs which can be found in nonverbal animal behavior. The difference which she sees between the signs of animals and those of human language is that the former are genetically replicated whereas the latter are replicated by learning, that is, in the form of sign tokens (re)produced from "some earlier token or tokens of its type that the speaker has heard" (Millikan, 1984, p. 22). In the long process of their replication, cultural signs "never lose their character of being also natural signs (Millikan, 2004, p. 105). Like the nonverbal signs of animals, words and sentence patterns survive as sign types by being reproduced in the form of their tokens (Millikan, 2005, p. 3). All cultural signs have their own "reasons for survival" (1984, p. 28). A major purpose of natural and cultural signs is to replicate in their usage. More specific reasons for the survival of signs are described by Millikan as follows:

> Artifacts that have been serving certain functions known to those who reproduce them and that are reproduced on this account (e.g., household screwdrivers) have these functions as direct proper functions. Behaviors that result from training or from trial and error learning involving correlations of a reward with the behavior have as direct proper functions to produce that reward. Behaviors that result from imitation of behaviors of others because the latter behaviors have correlated, within the observation of the learner, with certain functions have these functions as direct proper functions. (Millikan, 1984, p. 28)

The final causes for the replication of words and syntactic patterns are the effects which utterances evince in successfully coordinated acts of verbal communication (cf. Millikan, 2005, p. 45). "Conventional language devices are selected for performing services satisfactorily at once to both partners in communication" (2004, p. 26) since "a primary function of the human language faculty is to support communication" (2005, p. 39). The form of a speech act that succeeds in implanting the speaker's belief in the hearer's mind is "selected for reproduction" and is replicated (2005, p. 39). The grammatical form of the indicative mood, for example, survives because of the success which sentences in the indicative mood, in contrast to the ones in the subjunctive or imperative mood, have. Sentences in the indicative mood are verbal forms serving successfully as "guides in forming beliefs"; without the success in the use of this grammatical form, "speakers would stop trying to use them to impart beliefs. Production of true beliefs, then, is a linguistic function, a purpose, of the indicative form itself, whether or not a particular speaker and/or hearer have as their own purpose to use it in that way on a given occasion" (2005, pp. 190-191). Millikan

(2004, pp. 14-27) does not fail to acknowledge her indebtedness to Dawkins's theory of *memes*, which postulates that cultural products, such as words, catch phrases, slogans, or melodies, spread in human minds like genes spread in living species (Dawkins, 1976).

The "proper purpose" which a sign has independently of its user is hence to survive and to proliferate by staying in circulation (Millikan, 2005, p. 53). Not restricted to a theory of verbal signs, Millikan's teleosemiotics is a *general* theory of signs (Millikan, 1984, p. 85), which explains the replication of genetically transmitted sign behavior of animals as well as the proliferation of artifacts in culture. The purpose of a *technological meme*, such as an arrow whose design is copied again and again because it shoots well, is self-replication "by serving people's prior interests" (Millikan, 2004, p. 18). Millikan gives the following evolutionary account of the main difference between signs and instruments: artifacts, such as tools and other instruments, have individual inventors, whose inventions, if successful, tends to be replicated; verbal signs, by contrast, evolve like species or body organs, which have no individual inventor but an ancestor whose genes have proliferated (1984, p. 3). Furthermore, there is a difference in the way they replicate. Whereas words are replicated from one or many *tokens* in the process of language learning (see above), instruments are signs which are always replicated from a *type*: "Items produced by mass production are usually reproductions of some prototype not produced on the line – some original experimental model" (1984, p. 21). Words are not replicated by means of genes, but from precedents and conventions "thought to have little tendency to emerge or reemerge in the absence of precedent" (2005, p. 7). Language conventions are "lineages of behavioral patterns" repeating precedents of successful communicative interactions (2005, p. 86).

A thorough analysis of the affinities and differences between Peirce's and Millikan's teleosemiotics is a desideratum of current semiotic theory, but only a few notes for a critique of Millikan's teleosemantics can be given here. In light of Peirce's much more comprehensive evolutionary theory of semiosis, Millikan's principles of teleosemantics must appear reductionistic in the following respects:

1. Millikan's concept of sign is much narrower that Peirce's theory of the sign. Her restriction to signs of human beings and signs in animal behavior reduces her semiotics to the field of anthroposemiotics and zoosemiotics, ignoring the broader implications of biosemiotics, the semiotics of nature and certainly the semiotics of machines.

2. Millikan's account of signs surviving as types reproduced in the form of tokens shows the restriction of her semiotics to symbolic legisigns and their replicas; her semiotics ignores the semiotics of qualisigns, sinsigns, iconic, and indexical signs.

3. Millikan's theory of semiotic teleology and her views concerning the purpose of signs are much narrower than Peirce's teleological theory of semiosis (see Short, 1983).

4. Millikan's indebtedness to Dawkins's theory of memes and her account of sign evolution as a process of sign replication (parallel to Dawkins' theory of meme replication) make her teleosemiotics liable to the same fundamental semiotic critique which has been addressed against Dawkins (Deacon, 1999; Kull, 2000): sign evolution is not restricted to copying, self-reproduction, and the replication of types; these are certainly aspects of evolution but unable to account for semiotic creativity and what Peirce calls "the growth of signs."

5. Whereas Millikan's teleosemiotics is restricted to the Darwinist biological paradigm of evolution in which self-replication is the only evolutionary purpose of genes, species and, by extension, signs, Peirce's paradigm is also the one of a logic of signs. Signs are not only self-replications in the infinite process of semiosis, they also develop in a process of semiotic growth: "The purpose of signs" says Peirce (CP 2.444, fn, c.1893), "—which is the purpose of thought—is to bring truth to expression."

3. Instruments as Signs and Their Practical Indexicality

To conclude with Peirce that signs are not instruments, or perhaps more than mere instruments, is not incompatible with the conclusion that tools or instruments are signs. For different reasons, Peirce, Wittgenstein, Millikan, cultural anthropology, and the semiotics of culture agree that they are. Tools are products and hence signs of the culture of which they are part and which they constitute according to the framework of the semiotics of culture. Both instruments and words have functions and are used for purposes by their users (Wittgenstein). A single tool is the token of a type, and all replicas of this type serve the same purpose (Millikan). As a sign, an instrument represents an object, which is our previous knowledge of its utility, form, and modes of usage; its interpretant is the particular use a tool user can make of it in the future, which can be a thought about how to use it where and when or a practical use of the tool in a particular act of working with it (Peirce). Tools, just like verbal signs, are cultural extensions of the human mind into its external environment (Logan, 2007).

In a discussion of the differences between signs and instruments in a biological context, Fernández (2008) comes to the opposite conclusion: Instruments are not signs. His line of argument is the following: "They differ, among other things *in the way they are embodied in the world*" because "signs act independently of the properties of their physical embodiment. A large token of the letter A made of wood, for instance, achieves the same semiotic results as a small token made of ink. Instrumental relationality, on the contrary, is essentially dependent on the properties and qualities of its embodiments" (Fernández, p. 353). Fernández's arguments are not convincing. Firstly, instruments as tokens can differ as much from other instruments of the same type as one letter of the alphabet can differ from another. There are as many forms, colors, and styles of hammers as there are ways of pronouncing one and the same word, but despite the many materials from which hammers may be produced, all hammers serve the same practical purpose. Secondly, what Fernández says about the independence of the sign from the object which it represents is only true of arbitrary symbols.

It is true that tools, considered as signs, are not arbitrary symbols since their form is motivated by their practical utility, but this does not preclude that they may not be symbols for other reasons. The first and foremost criterion of the symbolic sign, according to Peirce, is not its arbitrariness, but its being a sign because of a habit which relates the sign to its object (cf. Nöth, in press). Tools are indeed cultural symbols for this reason, but in addition to being cultural symbols, tools are also motivated signs. The relation between the tool-sign and its object, which is its practical utility, is one of physical necessity. Only good tools serve their purpose; their quality depends on natural causes, which means that tools from this point of view are indexical signs. Screwdrivers and wrenches, for example, are designed at their ends in a way that their form fits the screws and bolts which to turn is their purpose; their handles must fit the human hand. In their mediation between the human hand and the type of screw or bolt for which they are designed they are thus twice motivated in their form by relations of practical complementarity.

The insight that instruments are motivated signs is also implicit in an argument made in Plato's *Cratylus*, in which Socrates expounds that the craftsman's instruments must be "right and natural" to the purpose they serve. In these words, Socrates describes the relation of complementarity between the instrument and the purpose it serves. If the instrument is considered as a sign, and the purpose as its object, the sign, in Socrates' description, is a natural sign. Today, natural signs are mostly considered to be indexical signs, but in earlier centuries of the history of semiotics, natural signs also comprised the class of iconic signs (cf. Nöth, 2008a). When he emphasizes that the craftsman's instruments must be "right and natural" to the purpose they serve, Socrates implies that these instruments should evince a natural correspondence of complementarity with the objects on which they operate in their environment. His message is that tools as well as signs must correspond to their objects, like an organism corresponds to its *umwelt* or *wirkwelt* according to von Uexküll (1940), or human and animal visual cognition corresponds in a complementary way to the ecological *affordances* of the objects in their environment according to Gibson's ecological psychology (Gibson, 1979; Lombardo, 1987).

Recently, parallels in the semiotic complementarity of the sign – object relationship between verbal signs on the one hand and tools on the other hand have been studied in the evolutionary perspectives of the theory of embodied cognition. In this framework, Andy Clark sees the following parallels between the sign nature of tools and language:

> Consider a familiar tool or artifact, say a pair of scissors. Such an artifact typically exhibits a kind of double adaptation – a two-way fit, both to the user and to the task. On the one hand, the shape of the scissors is remarkably well fitted to the form and the manipulative capacities of the human hand. On the other hand (so to speak), the artifact, when it is in use, confers on the agent some characteristic powers or capacities which humans do not naturally possess: the ability to make neat straight cuts in certain papers and fabrics … .

Public language is in many ways the ultimate artifact. Not only does it confer on us added powers of communication; it also enables us to reshape a variety of difficult but important tasks into formats better suited to the basic computational capacities of the human brain. Just as scissors enable us to exploit our basic manipulative capacities to fulfill new ends, language enables us to exploit capacities of pattern cognition and transformation in ways that reach out to new behavioral and intellectual horizons. ... The sheer intimacy of the relations between human thought and the tools of public language bequeaths an interesting puzzle ...; it is a delicate matter to determine where the user ends and the tool begins! (Clark, 1997, pp. 193-194)

Clark highlights and illustrates very well the relationship of necessary complementarity between tools and their use, which constitutes tools as indexical signs of a use determined by physical necessity. He is also right in concluding that tools as well as verbal signs are extensions of the human mind. However, his premise that "language is a tool" (Clark, p. 193) is reductionistic since it does not do justices to the aspect of the autonomy of verbal signs and their agency in the process of semiosis. It ignores the essential difference that tools are determined by their practical utility which makes them natural indexical signs of this utility. Language, by contrast, consists of signs which represent their objects in essentially symbolic ways. Above all, the agency of tools is one of efficient causality, whereas the agency of signs is one of final causality.

4. Semiotic Machines: Instrumental Replicators or Autonomous Agents?

The general context of the instrumental theory of the sign and the semiotics of tools suggests new answers to the question in which sense computers and robots are semiotic machines (Nöth, 2001a, 2001b, 2002, 2008b). According to the above premises, we can conclude that semiotic machines are not only semiotic instruments and instrumental replicators of signs, but also signs replicating themselves as signs. What remains to be examined is the question whether and in which respect semiotic machines can operate as autonomous semiotic agents.

4.1 Trivial Sign Processing Machines: Heteronomous Agency and Efficient Semiotic Causality

Semiotic machines differ as to their degrees of complexity and their semiotic potential. Foerster (1993) distinguishes *trivial* from *nontrivial machines*. Trivial machines, also called deterministic machines, operate by mere efficient causality since their output is completely determined by their input. The semiotic agency of these machines is heteronomous since their goals are determined by the purposes of operators and programmers, which are external agents from the point of view of the machine. Semiotic machines of this kind serve as mere instruments without any semiotic agency of their own. Before discussing the semiotics of nontrivial machines (see 4.3), let us first examine the mode of sign processing of trivial digital machines.

Copy machines, calculators, word processors, and other relatively trivial semiotic machines replicate signs, but do so by heteronomous agency. It was Flusser who introduced the concept of *sign producing instrument* as a fundamental concept of the history of the media. Is a copy machine a sign producing instrument? In his philosophy of photography, Flusser (1997) draws the distinction between simple tools and machines on the one hand and apparatuses on the other. His definition of apparatus in a sense restricted to the media is: "a tool for the production of techno-pictures" (Flusser, 1996, p. 150). The prototype of Flusser's apparatuses in this sense is the camera (German: *Fotoapparat*). Whereas the most general characteristic of tools and machines is that they perform *work* by *informing* in the sense of transforming the world, the purpose of an apparatus is to produce *signs* (which Flusser calls symbols; 1997, p. 23). What Flusser describes as a revolutionary step in the history of the media, that is, the invention of *sign producing machines*, has many precursors in the cultural history of tools. Thermometers, sundials, meter rules, and astronomical instruments are also sign producing instruments in this sense. What they have in common with the camera is that they produce natural indexical signs by processes of efficient causation. The signs that a camera produces are natural indices insofar as its pictures are the natural effect of the projection of light rays on a film. Natural signs have no sign producer and therefore, they cannot deceive, although they can be used to deceive (Nöth, 1997).

However, on the premises of Peirce's semiotics, it is as inadequate to call a camera a sign producing apparatus and thus to presuppose that it produces signs as it is inadequate to say that a sundial produces signs. Sign production presupposes autonomous semiotic agency, but the signs which a camera "produces" by its proper agency (disregarding the photographer who is indeed a sign producer) are natural signs, which cannot be said to be produced because natural signs have no sign producer. What these semiotic instruments do should be better called natural *sign processing*, with the term *processing* taken in the sense of natural transformation. In contrast to sign production, sign processing is also more suitable because it avoids the risk of the inappropriate assumption that a sign can be produced from something that is not a sign. With Peirce, we have to assume, instead, that an apparatus, such as the camera, does not *produce* a sign, at least not from something that is not a sign, but that it *transforms* one natural sign, consisting of the light rays reflected by an illuminated object, into another natural sign, which is the exposed film or the photo developed from it. In sum, semiotic instruments and apparatuses do not produce, but they process signs.

With this modification, Flusser's theory of the apparatus can be fruitfully extended to the study of semiotic machines, such as calculators, computers, or robots, although Flusser's distinction between apparatuses and machines, too, requires a revision. Computers and robots are not only apparatuses processing signs; they are also machines performing work, as all other machines do. Insofar as they act instrumentally as co-agents of sign processing and replication, semiotic machines operate without any autonomous agency of their own. In this respect, trivial copy and text processing machines differ from their nondigital precursors, such as the manually operated duplicating machine or even the simple typewriter, only in their degree of efficiency. It is also inadequate to conclude with Flusser that in fundamental contrast to trivial nondigital machines, which serve to execute physical work, digital sign processing machines serve to execute mental work. The distinction between physical and mental work has too often been taken for fundamental; with the invention of robots, which serve to perform both mental and physical work, it has become obsolete (cf. Nöth, 2008b). Furthermore, the Cartesian dualism between mental and physical work has always failed to recognize that the work of the typist at a typewriter, the copyist in front of a manuscript, or the painter painting a picture is not only mental but also manual work, which requires manual skills and muscular effort, and that the manual laborer cannot exert muscular effort in the process of work without doing mental work at the same time.

In sum, trivial or deterministic sign processing machines, such as cameras, copying machines, calculators, or word processors, produce signs by efficient causality. The product given by an electronic calculator as its output after the input of several factors and the command to multiply them is an example of a sign produced by efficient causation; this output is produced as the effect of a chain of dyadic cause-effect sequences in which electronic impulses have electronic effects determined by rigid programs and physical necessity. In such processes of sign processing by efficient causation, the trivial semiotic machine does not act as an autonomous agent since its agency is not determined by the machine itself but by physical causes which have immediate physical effects.

A copying machine does not interpret signs; it produces signs by efficient causation, but the signs produced a copy machine are of two kinds. Firstly, there is the sign that is merely the reproduction of its original. This mere reproduction of an original sign is not a new sign; it is merely the replica of a sign which continues to act with the same purpose of creating the same interpretant. Secondly, the copy is an indexical sign of its original, which is its object and to which it is related by efficient causes. As such, it creates an interpretant of its own, such as the user's judgment that it is a good or bad, faithful or false copy. In relation to the copy as an indexical sign of its original, the machine operates as sign producing agent, but not entirely by its own autonomy, since its purpose, to produce copies is the purpose of its designers and users. Their purpose is merely to create excellent copies of originals, whatever kind of sign these original signs may be. Like all indexical signs, the copy qua natural sign of its original cannot lie.

4.2 Semiotic Agency in Semiotic Machines as Instruments and as Signs
The conclusions that trivial semiotic machines only process signs instrumentally by efficient causality and that their participation in the process of semiosis is heteronomous since not they, but their operators are the true sign producing agents, do not mean that the operators of semiotic machines are fully autonomous semiotic agents. The semiotic autonomy of an operator processing signs by means of semiotic machines is restricted for two reasons. One is the autonomy of the signs produced by their operator, which have a logic of their own that cannot be ignored, and hence determined, by sign users who want to use them for their purposes. The other

restriction of the autonomy of the operators of a semiotic machine is that their semiotic operations depend on the efficient agency of the machines without which they would be unable to produce and to replicate their signs as efficiently. Such restrictions are the reverse of the increase in the semiotic potential of semiotic signs brought about by each invention of a new semiotic instrument. The autonomy of sign producers has always been limited by the restrictions inherent in their semiotic instruments. Semiotic tools contributing to the process of semiosis by efficient causation are necessary for the purpose of semiosis even though their instrumental agency is not a sufficient cause of the process of semiosis.

The interdependencies between humans, their instruments, and their signs have always been such that the extensions of humans, from the early primitive tools of *homo habilis* (alias *faber*) to the intelligent semiotic machines of the age of the World Wide Web, evince a bidirectional causality to the effect that the extensions of humans determine the thus extended humans and vice versa. In the interface of humans with computers, the bidirectional interdependencies which have always existed between instruments and their operators have only become more apparent. The intelligent extensions of the human mind are not only the most advanced semiotic instruments at their operators' disposal, they do not only liberate their users from mental work, but also enslave them in certain respects, and thus evince a tendency to become their masters. Semiotic machines do not only facilitate calculating, thinking, arguing, and communicating, they also become co-authors of the writing and calculating processes for which they are used insofar as they oblige their operators to work with the restricted range of operations which they permit and, last but not least, force them to update their programs, protect them against viruses, and do other kind of maintenance work in the service of the very machines which pretend to be their servants.

All semiotic machines do not only process signs; as objects of culture, they are also signs by themselves in the sense that all instruments are signs (see section 3). As such, they evince their own autonomous semiotic agency. Like all other signs, they have the purpose of creating an interpretant, to replicate, and grow in cultural, technological, and economic evolution, not only according to the principle of the survival of the semiotically fittest. The agency of semiotic machines in this sense is not reduced to the processing of *other* signs by efficient causality. Furthermore, computers have become media (Bolz, Kittler & Tholen, 1994), and as media, they are signs (Santaella, 1998). The mediality of semiotic machines makes them signs in the sense epitomized by McLuhan: like all other media, semiotic machines have a content of their own which differs from the content of media preceding them. Like other signs which grow with their interpretants, semiotic machines extend the semiotic potential of their users.

4.3 Rudiments of Autonomous Semiotic Agency in Nontrivial Semiotic Machines
The hot debate in computer semiotics and philosophy is not about sign processing by efficient causality but about the question whether semiotic machines can also act by final causality, that is, whether and to which degree they act as autonomous agents

with purposes of their own in the process of semiosis. Rudiments of semiotic autonomy to be briefly reviewed in the following are the autonomy of the machine in determining its own goals, intentionality, and dialogic reasoning.

One of the symptoms of autonomy is that the agents can set their own goals. On this premise, Haselager arrives at the conclusion that traditional top-down programmed robots are not autonomous agents: "They may be operating independently—even 'freely' choosing how to act in order to achieve goals—but the goals they are trying to achieve are still set by human programmers" (Haselager, 2007, p. 65). At the same time, the author recognizes that autonomous agency is theoretically not unattainable in the evolution of robots. Actually, it is the goal of a research field called *(co-)evolutionary robotics*: to develop self-organizing robots able to optimize their own morphology and to code their own control system in co-evolutionary processes of selection in which "the genotypes of the selected robots are subjected to crossover with other genotypes of the selected robots and further random mutation, giving rise to a new generation of robots" (Haselager, p. 70). The project of designing robots which will co-evolve according to the "affordances", necessities, and restrictions of the environment in which they are situated is inspired by analogies with the evolution of biological organisms in their particular ecological niches. As Ziemke summarizes, the project "might be a first step to developing robotic agents with (some of) the integration and coherence of living organisms, by rooting them in their environments through co-evolution of robot bodies, control systems, and their environments" (Ziemke, 1999, p. 186). Robots evolving to semiotic autonomy in this way are more than mere instruments; they evince characteristics of the autonomy of living beings. On the other hand, it must also be said that the extent of semiotic autonomy of human beings has been much overrated. The freedom of humans to choose their own goals is restricted in many respects; genetic, cognitive, social, or legal constraints restrict this freedom of choice in many ways. Furthermore, humans too often act as instruments of other human beings. For example, when executing paid work, workers do not act as autonomous but as heteronomous agents to the degree that their work is determined by the goals set by their company.

To set one's own goals requires the intention to do so. Intentionality, by definition involving final causality, is probably the most frequently addressed general symptom of semiotic autonomy, and it has often been argued that computers and robots lack intentionality (Dreyfus, 1972; Searle, 1980). In this debate, intentionality is usually defined as the purpose of autonomous semiotic agents using signs as their instruments for the purpose of achieving their goals. On these premises, the autonomy of the sign user is overrated since the agency of the sign in the process of semiosis is ignored. Sign users are not autonomous agents to the degree that the restrictions imposed on them by the logic of the sign itself restrict their semiotic choices (cf. section 4.2). Furthermore, the debate about the lack of intentionality in semiotic machines suffers from a dualistic fallacy of *tertium non datur*, the assumption that systems either have or do not have intentionality. In the context of evolutionary robotics, it is currently being asked whether intentionality and even consciousness should not rather be conceived of as emergent properties in semiotic systems. The lack of elements of intentionality in currently available intelligent machines should not preclude the possibility of the emergence of intentionality and consciousness in future generations of robots (Zlatev, 2001).

The evolutionary transition from trivial semiotic machines operating merely by efficient causality (see section 4.1) to nontrivial semiotic machines with rudiments of semiotic autonomy began long ago with feedback control systems (cf. Nöth, 1977). Such machines have the capacity of self-control, which is certainly an element of semiotic autonomy. Computers evince the capacity of self-control in processes of mechanical autodiagnosis: the machine produces signs about its own internal states, such as a technical defect, lack of electricity, or storage capacity. When these signs are reported back to the machine's control center in feed-back circles, such machines evince another rudiment of autonomy, which is self-correction. Further developed self-correcting machines (cf. Carbonell, 1990) even testify to rudiments of metasemiosis in semiotic machines, since self-correction presupposes the interpretation of signs by or by means of signs.

These and other rudiments of semiotic autonomy are the characteristics of *nontrivial machines* (Foerster, 1993), which do not merely operate by efficient causality. The output of a nontrivial machine is no longer completely determined by its input. The path to the achievement of a final cause is equally undetermined because it allows alternative routes. The prototype of a nontrivial semiotic machine is Brooks's (1991) bottom-up robot, which learns to interact with its environment from mere trial and error without being preprogrammed with representations of the world with which they interact. Instead of having such representation implemented in their own control center, these nontrivial semiotic machines construct the representations of the world with which they interact by learning from trial and error and memorizing the results from interacting with their environment.

The manifold technological opportunities of implementing more and more features of autonomous agency must not block insights into the remaining impediments on the way towards the rise of genuinely autonomous semiotic machines. Only three of them can be mentioned here: autopoiesis, self-consciousness, and genuinely dialogical semiosis.

Despite their rudimentary potential to set their own goals, semiotic machines will most likely remain heteronomous in essential respects. At least, the ultimate goals of their existence will continue to be set by their constructors who determine them according to the purposes of those who want to use them as their instruments. A machine that determines its own ultimate goals is either a machine replicating itself by its own agency (autopoiesis) or one that changes the ultimate goals set by its original constructor. Although research in artificial life has shown that machines with rudiments of autopoiesis are at least theoretically possible, machines that create their own offspring have not yet become reality. Even if machines were to self-replicate themselves one day, giving rise to artificial life, their autopoietic action would be restricted to replicating themselves in forms identical with themselves so that such processes of identical self-replication would lack genetic growth as it occurs in the natural evolution of biological organisms (cf. Nöth, 2002).

As far as the possibility of a machine able to change its ultimate goals is concerned, that is, a machine determining the ultimate reason for its existence, the construction of such a machine is not even desirable, since this machine could once decide to serve no longer as an instrument of its users but become their true masters. Similar arguments hold for the perspective of machines with genuine self-consciousness. Truly self-conscious agents are fully responsible for their actions so that a machine which would assume the full responsibility for its agency would become ethically responsible for its acts. Self-consciousness would make them social beings subject to social rewards and sanctions, praise for the good, and blame or even punishment for the bad achievements of its self-chosen goals.

Will semiotic machines ever operate in processes of semiosis involving creative inner dialogues? It was Peirce who raised this question as a criterion to examine the degree to which the reasoning capacity of a machine may become as highly developed as that of human beings. Against the background of the logical machines of his time, which he was willing to dub *reasoning machines*, Peirce concluded that such machines, unlike human beings, do not operate in creative inner dialogues in which one "critical self ... is trying to persuade" another self "that is just coming into life" in this very process (CP 5.421, 1905). Despite the rudiments of dialogicity in computers in their dialogical interaction with their users and in processes of weighing the alternatives of operating in one way or another at a given moment, it is unlikely that semiotic machines will ever be programmed to carry out creative inner dialogues. Machines with a self to be dialogically persuaded and even less so, machines with a truly critical self are not desirable, not least since inner dialogues can lead to self-doubt which may end in self-destruction and the loss of the machine.

4.4 Situatedness and Embodiment of Semiotic Machines

Peirce recognized that the logical machines of his time were restricted in their semiotic potential for another reason, the lack of practical experience of, and interaction with, their umwelt. Natural reasoning is more than the mere processing of coded signs and their combination according to syntactic rules; it requires the "collateral experience" (CP 8.179, 1903) of the world which the reasoning is about, since natural "reasoning is not done by the unaided brain, but needs the cooperation of the eyes and the hands," as Peirce wrote in 1887 (quoted in Ketner & Stewart, 1984, pp. 208-209).

Peirce's insight into the inadequacy of reasoning machines which lack collateral world experience anticipated insights of modern research in artificial intelligence, cognitive science, cybersemiotics (Brier, 2008), and robot semiotics currently discussed as situatedness, cognitive grounding, embodiment (Ziemke & Sharkey, 2001), and *enaction* (Varela, Thompson, & Rosch, 1991) whose concern is the reciprocity between the semiotic agency of organisms and machines on the one hand and their environment on the other. It is fundamental to the growing insight of cognitive science that the human mind is an *extended mind* (Logan, 2007) whose expanse reaches beyond the confines of the human brain and comprises also the external signs which human minds have produced and left to be interpreted in their external environment in the form of writing, images, instruments, and all other semiotic manifestations of culture.

In comparison to the current theories of the extended mind, of the cognitive complementarity of semiotic agents and the environment with which they interact, partially inspired by Uexküll's theory of the interdependencies between the organism's *innenwelt* and its *umwelt* (Ziemke & Sharkey, 2001), Peirce's theory of semiosis is more ambitious in its scope. Peirce's evolutionary semiotics does not only look at the way the human mind extends itself semiotically into its own environment. Instead, it also ascribes autonomy, rationality, and even intelligence to the signs outside the human brain in the material world as well as to the machines created by human beings as their instruments. Signs, according to Peirce, are not only the products of semiotic agency; they are endowed with an agency of their own. Therefore, rationality and intelligence are not only characteristics of human minds and their semiotic extensions, but they can also be found in the material world. According to Peirce, natural human reasoning and the artificial processes of reasoning in lifeless machines have common evolutionary roots. In the evolution of the universe from matter to mind, the material and the living world have evolved under the same physical constraints, so that the processes of reasoning in human thought operates, at least at the level at which mathematical and elementary logical operations are concerned, in fundamentally analogous ways (cf. Nöth, 2004).

An important mode of sign processing common to humans and machines is diagrammatic iconicity. Peirce (1887, p. 168) states: "The secret of all reasoning machines is after all very simple. It is that whatever relation among the objects reasoned about is destined to be the hinge of a ratiocination, that same general relation must be capable of being introduced between certain parts of the machine." Notice that a precursor of this lock-and-key argument of the diagrammatic correspondence between the way machines operate and the structures of the world which their operations represent is Socrates's account of the relationship between the craftsman's tool and the purpose for which it serves. When Socrates emphasizes that the craftsman's instrument must be "right and natural" to the purpose it serves, he says that any good instrument should be a diagrammatic icon with characteristics corresponding to the one of its field of operation.

References

Bolz, N., Kittler, F. & Tholen, G. C. (1994). *Computer als Medium*. Munich: Fink.
Brier, S. (2008). *Cybersemiotics: Why information is not enough*. Toronto: Toronto University Press.
Brooks, R. A. (1991). Intelligence without representation. *Artificial Intelligence*, *47*, 139-159.
Bühler, K. (1934). *Sprachtheorie*. Stuttgart: Fischer. (re-edition, 1965)
Carbonell, J. (Ed.) (1990). *Machine learning*. Cambridge, MA: The MIT Press.
Carruthers, P. (1996). *Language, thought, and consciousness*. Cambridge: Cambridge University Press.
Clark, A. (1997). *Being there: Putting brain, body, and world together Again*. Cambridge, MA: The MIT Press.
Dawkins, R. (1989). *The selfish gene* (rev. ed.). Oxford: Oxford University Press. (first published in 1976)

Deacon, T. (1997). *The symbolic species*. New York: Norton.
Deacon, T. (1999). Memes as signs. *The Semiotic Review of Books*, *10* (3), 1-3.
Deely, J. (2001). *Four ages of understanding*. Toronto: University of Toronto Press.
Donald, M. (2001). *Origins of modern mind*. Cambridge, MA: Harvard University Press.
Dretske, F. (1983). *The nature of psychological explanation*. Cambridge, MA: The MIT Press.
Dretske, F. (1988). *Explaining behavior*. Cambridge, MA: The MIT Press.
Dreyfus, H. (1972). *What computers (still) can't do: A critique of artificial reason*. Cambridge, MA: The MIT Press.
Fernández, E. (2008). Signs and instruments: The convergence of Aristotelian and Kantian intuitions in biosemiotics. *Biosemiotics*, *1*, 347-359.
Flusser, V. (1997). *Für eine Philosophie der Fotografie* (8th ed.). Göttingen: European Photography.
Flusser, V. (1998). *Kommunikologie* (rev. ed.). Frankfurt/Main: Fischer.
Fodor, J. (1975). *The language of thought*. Cambridge, MA: Harvard University Press.
Foerster, H. v. (1993). *Wissen und Gewissen: Versuch einer Brücke*, (S. J. Schmidt, Ed.). Frankfurt/Main: Suhrkamp.
Gibson, J. J. (1979). *The ecological approach to visual perception*. Hillsdale, NJ: Erlbaum.
Haselager, W. F. G. (2007). Robotics, philosophy and the problem of autonomy. In I. E. Dror (Ed.), *Cognitive technologies and the pragmatics of cognition* (pp. 61-77). Amsterdam: Benjamins.
Kaku, M. (1997). *Visions: How science will revolutionize the 21st century*. New York: Anchor Books.
Kappner, S. (2004). *Intensionalität in semiotischer Sicht: Peirceanische Perspektiven*. Berlin: de Gruyter.
Keller, R. (1995). *Zeichentheorie*. Tübingen: Francke. (English version is available: Keller, R. (1998). *A Theory of Linguistic Signs* [K. Duenwald, Trans.]. Oxford: University Press)
Ketner, K. L. & Stewart, A. F. (1984). The early history of computer design: C. S. Peirce and Marquand's logical machines. *Princeton University Library Chronicle*, *55* (3), 187-224.
Kull, K. (2000). Copy vs. tanslate, meme vs. sign: Development of biological textuality. *S. European Journal for Semiotic Studies*, *12* (1), 101-120.
Logan, R. T. (2007). *The extended mind: The emergence of language, the human mind, and culture*. Toronto: University of Toronto Press.
Lombardo, T. J. (1987). *The reciprocity of perceiver and environment*. Hillsdale, NJ: Erlbaum.
McLuhan, M. (1964). *Understanding media: The extensions of man*. London: Routledge & Kegan.
Meier-Oeser, S. (1997). *Die Spur des Zeichens: Das Zeichen und seine Funktion in der Philosophie des Mittelalters und der frühen Neuzeit*. Berlin: de Gruyter.
Millikan, R. (1984). *Language, thought, and other biological categories: New foundations for realism*. Cambridge, MA: The MIT Press.
Millikan, R. (2004). *Varieties of meaning*. Cambridge, MA: The MIT Press.
Millikan, R. (2005). *Language: A biological model*. Oxford: Clarendon.
MS: See Peirce (1979)
Neander, K. (1995). Misrepresenting and malfunctioning. *Philosophical Studies*, *79*, 109-141.
Nöth, W. (1977). *Dynamik semiotischer Systeme*. Stuttgart: Metzler.
Nöth, W. (1997). Can pictures lie? In W. Nöth (Ed.), *Semiotics of the media* (pp. 133-146). Berlin: Mouton de Gruyter.
Nöth, W. (2000). *Handbuch der Semiotik* (2nd rev. ed.). Stuttgart: Metzler.
Nöth, W. (2001a). Protosemiotics and physicosemiotics. In W. Nöth & K. Kull (Eds.), *Semiotics of Nature* [Special Issue]. *Σημειωτική: Sign Systems Studies*, *29* (1), 13-26.
Nöth, W. (2001b). Semiosis and the umwelt of a robot. *Semiotica*, *134*, 695-699.
Nöth, W. (2002). Semiotic machines. *Cybernetics & Human Knowing*, *9* (1), 5-22. (Also in P. Cobley [Ed.], 2006. *Communication theories: Critical concepts in media and cultural studies*, [vol. 4, pp. 341-360]. London: Routledge.)
Nöth, W. (2004). Semiogenesis in the evolution from nature to culture. In M. Alač & P. Violi (Eds.). *In the beginning: Origins of semiosis* (pp. 69-82). Turnhout, Netherlands: Brepols.
Nöth, W. (2007). Self-reference in the media. In W. Nöth & N. Bishara (Eds.), *Self-Reference in the media* (pp. 3-30). Berlin: Mouton de Gruyter.
Nöth, W. (2008a). Natural signs from a synechist perspective. In K.-M. Hingst & M. Liatsi (Eds.), *Pragmata. Festschrift für Klaus Oehler zum 80. Geburtstag* (pp. 130-140). Tübingen: Narr.
Nöth, W. (2008b). Sign machines in the framework of *Semiotics Unbounded. Semiotica*, *169*, 319-341.
Nöth, W. (in press). The criterion of habit in Peirce's definitions of the symbol. Transactions of the Charles S. Peirce Society.
Pagee, S. (1967). Of words and tools. *Inquiry*, *10*, 181-195.
Pape, H. (1993). Final causality in Peirce's semiotics and his classification of the sciences. *Transactions of the Charles S. Peirce Society*, *29* (4), 581-607.
Papineau, D. (1987). *Reality and representation*. Oxford: Blackwell.
Parmentier, R. J. (1985). Signs' place in medias res: Peirce's concept of semiotic mediation. In E. Mertz & R. J. Parmentier (Eds.), *Semiotic mediation* (pp. 23-48). Orlando, FL: Academic Press.
Peirce, C. S. (1887). Logical machines. *American Journal of Psychology*, *1* (1), 165-170. (Also in C. S. Peirce (2000). *Writings*, vol. 6, [N. Houser et al., Eds.; pp. 65-72]. Bloomimgton, IN: Indiana University Press)
Peirce, C. S. (1931–1935, & 1958). *The collected papers of Charles Sanders Peirce*. Vols. I–VI [C. Hartshorne & P. Weiss, Eds., 1931–1935], Vols. VII–VIII [A. W. Burks, Ed., 1958]. Cambridge, MA: Harvard University Press. (Citations use the common form: CP vol.paragraph)

Peirce, C. S. (1979). *The Charles S. Peirce Papers* (30 reels, 3rd microfilm ed.). Cambridge, MA: The Houghton Library, Harvard University, Microreproduction Service (citations use the form MS #, date).

Pinker, S. (1999). *How the mind works*. New York: Norton.

Plato. (1953). *The dialogues of Plato* (4th ed., B. Jowett, Trans.). Oxford: Clarendon.

Prieto, L. J. (1966). *Messages et signaux*. Paris: Presses Universitaires.

Prieto, L. J. (1972). *Nachrichten und Signale* (G. Wotjak, Trans.). Berlin: Akademie Verlag & München: Hueber. (Originally published as *Messages et signaux* by Presses Universitaires of Paris, 1966)

Prieto, L. J. (1973). Signe et instrument. In *Littérature, histoire linguistique: Recueil d'études offert à Bernard Gagnebin* (pp. 153-161). Lausanne: L'Age d'Homme.

Reddy, M. J. (1993). The conduit metaphor. In A. Ortony (Ed.). *Metaphor and thought* (2nd ed., pp. 164-201), Cambridge: Cambridge University Press. (first published in 1979)

Rolf, E. (2008). *Sprachtheorien. Von Saussure bis Millikan*. Berlin: de Gruyter.

Rossi-Landi, F. (1974). Linguistics and economics. In T. A. Sebeok (Ed.). *Current trends in linguistics: Vol. 12. Linguistics and adjacent arts and sciences* (pp. 1787-2017). The Hague: Mouton. (Was also published as a book: Rossi-Landi, F. (1975). *Linguistics and economics*. The Hague: Mouton)

Santaella, L. (1998). Der Computer als ein semiotisches Medium. In W. Nöth (Ed.). *Medientheorie und die digitalen Medien* (pp. 121-158). Kassel: Kassel University Press.

Santaella, L. (1999). A new causality for understanding the living. *Semiotica*, *127*, 497-519.

Saussure, F. de (1959). *Course in general linguistics* (W. Baskin, Trans.). New York: McGraw-Hill. (Originally published as *Cours de linguistique générale* by Payot of Paris, 1916)

Searle, J. (1980). Minds, brains, and programs. *Behavioral and Brain Sciences*, *3*, 417-424.

Short, T. L. (1983). Teleology in nature. *American Philosophical Quarterly*, *20*, 311-320.

Short, T. L. (2007). *Peirce's theory of signs*. Cambridge: Cambridge University Press.

Uexküll, J. von (1940). *Bedeutungslehre*. Frankfurt: Fischer.

Varela, F., Thompson, E. & Rosch, E. (1991). *The embodied mind: Cognitive science and human experience*. Cambridge, MA: The MIT Press.

Vygotsky, L. S. (1981). The instrumental method in psychology. In J. V. Wertsch (Ed.). *The concept of activity in Soviet psychology* (pp. 134-143). Armonk, NY: Sharpe. (Originally published in Russian in 1930)

Weinrich, H. (1958). Münze und Wort. In H. Lausberg & H. Weinrich (Eds.). *Romanica: Festschrift für Gerhard Rolfs* (pp. 508-521). Halle: Niemeyer.

Wittgenstein, L. (1953). *Philosophische Untersuchungen – Philosophical investigations* (Bilingual ed., G. E. M. Anscombe, Trans.). Oxford: Blackwell.

Ziemke, T. (1999). Rethinking grounding. In A. Riegler, M. Peschl & A. von Stein (Eds.). *Understanding representation in the cognitive sciences* (pp. 177-190). New York: Kluwer/Plenum.

Ziemke, T. & Sharkey, N. E. (2001). A stroll though the worlds of robots and animals: Applying Jakob von Uexküll's theory of meaning to adaptive robots and artificial life. *Semiotica*, *134*, 701-146.

Zlatev, J. (2001). The epigenesis of meaning in human beings, and possibly in robots. *Minds and Machines*, *11*, 155-195.

Craft, D. (2009). *The Ghosts of Future Brides*. Photomontage: Triple Golden Rectangles; 21 x 102 cm.

Cybernetics and Human Knowing. Vol. 16, nos. 3-4, pp. 37-79

The Grammar of Linguistic Semiotics
Reading Peirce in a Modern Linguistic Light

Per Durst-Andersen[1]

The paper presents a new typology of linguistic signs primarily based on Peirce's sign conception. It is demonstrated that the fundamental simple sign, the symbolic nominal lexeme, has an arbitrary relationship to its object in order to make it omnipotent, that is, open to various possible interpretations and therefore capable of referring to anything within its own limits. Exactly because of the arbitrariness and the omnipotence of the repertoire of nominal lexemes language must have a grammar that can give a semiotic direction to the otherwise completely static sign (in Peircean terms: impotent)—it cannot by itself refer, but must have a vehicle, a grammar. There is an obligatory choice between three types of grammar corresponding to the three ways in which a state of affairs exists in a communication situation: (1) the situation as such in a real or in an imagined world being common to the speaker and the hearer; (2) the speaker's own experience of the situation; or (3) the hearer's own experience of it. This means that the fundamental complex sign, namely the sentence, is first of all an index; it is a prime index—it is either (1) a model of the situation, (2) a symptom of the speaker's experience of it or (3) a signal to the hearer to find the situation behind the speaker's message. However, the individual grammatical signs found in a sentence have a double function. Besides their integrative function according to which the grammatical signs of dichotomic categories perform the role of an index, that is, they point to a goal, for instance, to the hearer, the individual members of the category perform the role of an icon, or its negative counterpart, in their differentiating function: They indicate whether or not there is a sign of equality, for instance, between the speaker's experience and that of the hearer. If there is a sign of equality— their experiences match—then I shall argue that the goal pointed to by the semiotic direction is reached. If there is no sign of equality—their experiences do not match—then I shall argue that the goal is not reached. We will stay in the speaker, because it is impossible to find a matching experience in the hearer's store of experiences. In other words, the static linguistic symbol needs an index that needs an icon (or the lack of an icon) in order for it to be properly decoded by the hearer. Furthermore, it is demonstrated that in addition to the level of naming objects (ensured by nouns) and situations (ensured by the verb)—the latter corresponding to Peirce's rhematic sign— and in addition to the level of assertion—corresponding to Peirce's dicentic sign—there is a third level at which verbal categories collaborate in order to make a deduction, abduction or induction— corresponding to Peirce's argumentative signs.

Key words: dichotomy vs. trichotomy; icon, index, symbol; arbitrariness; indexicality, diagrammatic relations; code; types of languages or grammar; situation, experience, information; deduction, abduction and induction; Peirce, Saussure, Bühler, Bakhtin and Jakobson

1. Introduction

Saussure considered the linguistic sign to be two-sided: consisting of a sound-image, the expression of the sign; and a concept, its content. The relationship between what he called the signifier (*signifiant*) and the signified (*signifié*) was regarded as arbitrary

---

1. Department of International Studies of Culture and Communication, Copenhagen Business School, Dalgas Have 15, DK-2000 Frederiksberg, Denmark. Email: pd.ikk@cbs.dk

and established by convention (Saussure, 1916, 66ff). This conception has been the starting point for many general linguistic theories since. Language consists of words or lexemes which are traditionally called symbols. (For Saussure and Piaget symbol would correspond to what we will call icon.) Although Saussure devotes the second part of his book (pp. 141-192) to grammar, the specific semiotic status of grammatical morphemes, so-called grammemes (e.g., *a*, *the*, and *–ed* in English), is not touched upon and the specific role of grammar in the process of semiosis is not mentioned.

Many other linguistic traditions or schools have also regarded words as symbols, and they too have only little to say about the semiotic status of grammatical entities and their specific role in semiosis. This concerns several varieties of old structuralism such as the Prague School (cf. Jakobson, 1971), the Copenhagen School (cf. Hjelmslev, 1943), the Russian tradition (cf. Serebrennikov, 1973), the Tartu School (cf. Lucid, 1977) and the American structuralist tradition from Bloomfield (1935) to Hockett (1958). It certainly concerns the different varieties of neostructuralism such as transformational grammar (Chomsky, 1957) and its various offshoots in the sixties (e.g., generative semantics) and seventies (e.g., space grammar), Government and Binding and other formal grammars (cf. Chomsky, 1981, 1986), Functional grammar (cf. Givón, 1981; Dik, 1989; Hengeveld, 2004) and cognitive grammar (cf. Langacker, 1999; Lakoff, 1987; Talmy, 2001). The difference between them is that the old structuralist tradition paid a lot of attention to the phonological and morphological levels of language, while the new structuralist tradition lost that interest, at least as far as synchronic studies are concerned, and instead concentrated their attention on syntax alone or its relations to semantics. In short, the interest went from the *simple sign* to what I call the *complex sign* (cf. Durst-Andersen, 2008) without discussing the semiotic consequences of doing so.

As a discipline linguistic semiotics has been more or less dormant since Jakobson died in 1982. This does not mean that linguists have not been dealing with language from a semiotic point of view—they surely have, for instance, Andersen (2001) and Shapiro (1983, 2003) both being pupils of Jakobson and inspired by Peirce, scholars belonging to the so-called Columbia School (also inspired by Jakobson; see, e.g., Contini-Morava et al., 2004) and Langacker (1999), just to mention some of the most important ones. However, works of this kind are contributions to linguistics made from a semiotic-based perspective. They are not contributions to the subfield of semiotics, called linguistic semiotics, in which language in its totality is regarded as a specific sign system compared to other sign systems. This discipline almost fell into oblivion with Jakobson. The main reason for that has to be found in the shift of interest in the beginning of the 1960s and that time's focus on Morris' three dimensions of semiosis, namely: syntax, semantics and pragmatics (cf. Morris, 1938). These three dimensions have indeed been studied intensively (the last 50 years bear witness to this), but the study of the complex sign itself and the semiosis process has been completely ignored. The result is that today's linguistics, from a purely semiotic point of view, presents itself in a very poor state and cannot be said to have developed any further since Saussure, Hjelmslev and Jakobson.

From this perspective, nobody has so far tried to apply Peirce's sign conception and his trichotomic system of thinking to the various semiotic entities of language, that is, to nominal and verbal lexemes, nominal and verbal grammemes, sentences, paragraphs and texts. This concerns Jakobson, too, despite the fact that he was attracted to Peirce and did not miss an opportunity to mention Peirce—especially when he was criticizing Saussure (see, e.g., Jakobson, 1962, 1963, 1965). But Jakobson never used Peirce in his own research. This is perhaps linked to Peirce's thrichotomic way of thinking which must have been difficult for Jakobson to accommodate to his own dichotomic way of thinking (or binary hocus pocus, as Wierzbicka (1980) puts it in the preface to her monograph). As I see it, there need not be any inconsistency in applying them in combination: dichotomies make perfect sense when making an analysis or performing a deconstruction of a mental building (that is why speech perception is a question of plus or minus), while trichotomies seem to make sense when performing a synthesis or a construction of a mental building (a symbol, for instance, is built upon an index that is built upon an icon).

The introduction of Peirce's sign conception to linguistic semiotics will automatically bring dynamism into a field that has been characterized by the opposite, presumably because of the focus on the language system and the simple sign. However, Peirce is, unfortunately, not enough despite the fact that he dealt with almost any subject and provided us with an enormous theoretical framework. We need other scientific inspirations, because Peirce died before the scientific boom in the 1920s and 1930s, when we find important contributions by Bühler, Bakhtin, and Jakobson. A theory of linguistics semiotics has, of course, to include a linguistic-based theory of communication. Here we take the starting point in Bühler's *Organonmodell* from 1934, but, as will appear from section 3, it is in no way a general communication model, although it has been treated as such—it only applies to the hearer and does not seem to include the function of grammar in communication for which the speaker alone is responsible. The specific hearer-orientation is not only true of Bühler's model, but is generally true of communication science (cf., e.g., relevance theory, Sperber & Wilson, 1986, 2004). Communication science is almost exclusively centred on the hearer, that is, the second person, whereby the speaker voice is ignored. To reach the level of a true dialogue and to bring the notion of dialogue into texts we have to include Bakhtin (1929/1994) and his important notion of voice and polyphony. Moreover, although Peirce, indeed, was a polyhistorian and a "pathfinder in linguistics" (Nöth, 2002, p. 1), and, indeed, made many interesting remarks on various levels of language as we shall witness in the following, he did not develop a complete theory of human language with a separate part on grammar. Therefore we have to find an approach to grammar that is compatible with Peirce's dynamism and process orientation. I find this in Jakobson's insistence on applying a teleological approach to grammatical categories: for what reason does this or that category exist? This, I think, will be in direct harmony with Peirce's general view.

What follows will be a short outline of a theory of linguistic semiotics that is the result of more than 25 years working on language and languages. It combines my own ideas with Peirce's semiotics, Bühler's communication theory, Bakhtin's notion of voice and his insistence on true dialogism, Hjelmslev's distinction between form and substance, Jakobson's functional or teleological linguistics, and the cognitive approach of modern linguistics. In other words, the present theory is meant as an original contribution to linguistic semiotics. It is not meant as a contribution to the enormous amount of research on Peirce's conception of the sign—I do not pretend to be a Peirce-specialist. I was brought up in the structuralist paradigm as it was practiced by Jakobson, by Hjelmslev, and by various Russian linguists (e.g., Mel'čuk, Gak, Klimov), but got inspired by Peirce, Bühler and Bakhtin who in many ways were ahead of their time. Such a mixture would normally yield a heterogeneous result, but, hopefully, not in this case.

2. The Simple Sign as a Static Unit – A Symbol

2.1. The Sign as a Three-sided Entity

The simple linguistic sign deprived of its grammemes, that is, grammatical prefixes, infixes and suffixes, is a so-called lexeme the function of which is to name something. A lexeme is traditionally considered to be a two-sided phenomenon: it has an expression side as well as a content side. Peirce, however, views any sign as a triadic phenomenon, namely the sign itself, sometimes also called the *representamen*, the *object* of the sign and its *interpretant*. The consequence of extending the number of sides from two to three is that the content of the linguistic sign, and not its expression, is divided into two different types. In other words, Peirce's object and interpretant have in common that they both designate some content, and that these two types of content are linked together by a representamen, the expression unit. In this specific context I cannot go into any detailed discussion of Peirce's conceptions (cf. Peirce, 1932, 1953), but there seems to be strong pieces of evidence for stating, on the one hand, that the (immediate) object is equivalent to what I shall argue is a mental image, that is an abstract pictorial-like representation in a person's mind, and, on the other hand, that the (immediate) interpretant is equivalent to what I shall argue is an idea, that is, a representation in a person's mind that is built upon abstract and generalized descriptions of the object derived from concrete thoughts of it (the only reservation is that feelings are treated by Peirce as belonging to the interpretant—I treat feelings like all the other four sense, i.e. as images). In short, a lexeme can be described as an image-idea pair.

If we base ourselves on Peirce's later writings where he gets far more specific (cf. Peirce, 1953, where his letters to Lady Welby are published), he does not only make a distinction between a mental image and idea, but also a sharp distinction between type and token more or less corresponding to what Hjelmslev called *form* and *substance*— a generalization of Saussure's distinction between *langue* and *parole*. But, when dealing with Peirce there is always a potential danger of having misunderstood him, because he was rich of words and often used words in a technical sense (this concerns, e.g., *dynamical* and *dynamically*). Moreover, from the point of view of today it is extremely difficult, if not impossible to distinguish between what Peirce actually meant and what he meant according to different Peirce scholars. Hopefully, I will not harm anyone when I say that one of the reasons for Peirce being so popular has to be found in the fact that his writings make room for scholars' individual conceptions. I do not exclude myself in that regard.

2.2. Nouns and Verbs as Different Signs

We have to make a sharp distinction between the image-idea pair of a common noun and that of a verb. As words both are symbols, but as to the structure and the content of their image-idea pairs they differ fundamentally. Although Peirce was aware of the difference between common nouns and verbs (see, e.g., the paragraph on syntax in Nöth, 2002), and indeed had many interesting things to say in this connection (see, e.g., CP 2.261, CP 3.459), he did not make a real sign distinction between them. Both were treated as being word-like in their naming functions and were subsumed under the notion of *rhemes* (for more about that, see sections 2.2.4 & 2.2.5). The reasons for making a sharp distinction are many, but I shall restrict myself to the most obvious.

2.2.1. The Mental Image of Common Nouns and Verbs

First of all, the mental image linked to a common noun consists either of a figure (*house, car, tree, elephant*) or a ground, be it heterogeneous (*golf course, tennis court, ice rink*), or homogeneous (*water, soil, sand, grain*). The mental image linked to a verb is always made up of a figure-ground constellation (or a corresponding part-whole construction or an element-set configuration), for example, *X be at L (ocation), X ride Y, X give Y to Z*. Moreover, some mental images will be stable, for example, *X be at L, X have Y, X need Y,* while others will be unstable, for example, *X work at L, X ride Y, X cry*. In addition to that, many verbs create not one, but two interrelated images, the one being unstable, the other being stable, for example, *X give Y to Z*, where <X is handing Y to Z> corresponds to the unstable image and <Z has Y> corresponds to the stable one. Verbs such as *X go to L; X bring Y to L, X borrow Y from Z, X tell Y to Z, X make Y something,* and so forth all create two interrelated images. Although it is possible to maintain that the mental image of *golf course* could contain the same elements as those of a certain verb construction, for example, *The flag is in the hole on green 18*, it is important to note that the former image will be unfocused, whereas the latter will be focused on the flag having the hole as ground and the surroundings as scene. It is also important to note that common nouns often create many mental images corresponding to the five senses: visual, auditory, gustatoty, olfactory and somatosensory images. Verbs tend to highlight one of them, for example, *see/look, hear/listen, taste, smell, feel*. In other words, verbs distinguish themselves from common nouns by being connected to either stable or unstable images, to single or double images, and last, but not least, by having a figure-ground constellation, that is, a certain perspective within a single domain.

2.2.2. The Idea of Common Nouns and Verbs

The idea linked to a verb is made up of a description of the figure-ground constellation in the mental image. The verb *X carry Y to Z* will create two images, an unstable where <X is carrying Y> as well as a stable one where <Z has Y>. This is described by what I call two ground-propositions: "While Y is sitting/hanging on X, X DO smth."(an activity description which entails a state description); and "Y BE with Z"(a state description).This has to be so, because when one says *She is carrying the child to her granny*, then the activity description is true, while the state description is false, whereas when one says that *She carried the child to her granny*, then the activity description as well as the state description are true. Thus, the two ground-propositions related to one another by the logical relation of implication make up the idea of the verb *X carry Y to Z*. Although the common noun *golf course* can be translated (as Peirce calls it) into "a location for rule-guided motion where one uses a club to make a ball run into a hole" which, indeed, is a proposition-based description or better: a definition, it is more likely that its idea-based description is made up of a hierarchical ordering of a closed set of descriptors that in each case is filled in according to what human beings know or may learn about the object named by the common noun *golf course*, for example, Category: artifact; Function: location for rule-guided motion; Subfunction: sport activity for human agents; Place: natural terrain; Instrument: club; Patient: small round ball; Goal: 18 small round holes placed at different places; Purpose of activity: make the ball run into the hole at Location 1b, 2b, … 18b by hitting it from Location 1a, 2a, … 18a and all necessary locations between 1a, 2b, … 18b and 1b, 2b, … 18b, respectively. This is what I call the idea of *golf course*. This is a structured description of what most people know about golf courses, from which it is possible to arrive at a precise definition. Definitions are, however, not made by ordinary people, but by scholars or others interested in nouns as technical terms.

2.2.3. Concluding Remarks

In short, based on the above-mentioned arguments I shall distinguish sharply between the sign structure of common nouns and that of verbs. Both nouns and verbs pair a mental image and an idea, but their structure is different. Such a distinction will also be more in accordance with how nouns and verbs are treated in linguistics. Verbs govern nouns according to their valency schema, whereas nouns may occupy various places in different valency schemata. Nouns are in this way subordinated to the verb. Another way of putting this is to say that the verb creates an image structure and a propositional structure in which nouns may take their place. As a matter of fact, Peirce saw the empty places given by a verbal lexeme long before linguists did. He said that predicates can be monadic, dyadic and triadic corresponding to what is called mono-, di- and trivalent in linguistics. But, nevertheless, he regarded both common nouns and verbs as *rhemes* (CP 2.252). I will reserve *rhemes* for verbs and use *terms* for common nouns (actually a notion that Peirce used before he substituted it for *rheme*).

2.3. Image vs. Idea and Form vs. Substance

Let me illustrate two important points by elaborating on a concrete example from Parker, a specialist on Peirce:

> a fresh footprint in mud is a representamen, its object is the person (or at least the foot) that created it, and its interpretant is the idea that someone recently walked in the spot. (Parker, 1997, p. 137)

Parker seems to concentrate on one thing, namely the concrete situation, that is, the substance itself, but, nevertheless, he is not specific, at all: Is the footprint made by a naked foot or by a shoe containing a foot? In my examination of this scenario I shall stick to the former, much simpler case.

Parker seems to forget that in order to be able to recognize a fresh footprint (sinsign—corresponding to token—in Peirce's terminology) as a footprint, the person in question should have an abstract representamen of a footprint which he uses as a kind of formula (a legisign—corresponding to type—in Peirce's terminology) when meeting all the different actualizations or replicae of it (so-called sinsigns) which involve various qualities or distinctive features (so-called qualisigns—corresponding to tone—in his terminology) (cf. CP 2.245-246):

Table 1: Representamen and Firstness, Secondness, and Thirdness

Representamen	Status	Exemplification
Qualisign	Tone	Possible Manifestations
Sinsign	Token	Fresh Footprint
Legisign	Type	Conventional Representation

We encounter the same in our daily life: we see different written versions of work, namely <labor> and <labour>, and hear different oral versions of it (rheme) depending on dialect, sociolect and idiolect, but we identify them all as being tokens (sinsigns) of the same type (legisign), because we have a (phonological) *form*, for example, /labʌr/, that can account for all possible realizations in the written as well as in the oral medium, including the one actually met at a particular point in time (legisign), for example, [leɪbə]. I shall argue that the form, or type in Peirce's terminology, manifests itself in many ways in the substance (or as tokens). Peirce would have said that the legisign determines the sinsigns. The expression form evoked by the specific footprint accounts for what sort of footprint the person in question identified, namely a human footprint and not a footprint from an ape.

The abstract expression recognized on the basis of the concrete expression, that is, the footprint in the mud, will simultaneously evoke a mental image (an object in Peirce's terminology) of a human foot, presumably viewed from above, because human feet are normally seen from that angle. I admit that the abstract expression could evoke a concrete picture, but in that case we would be dealing with an icon, that is, the foot and the footprint must necessarily belong to a person that is well-known to the observer. The mental image evoked will depend on the shape and seize of the footprint itself. It could be a prototypical girl's foot, boy's foot, any female or male foot. People have mental images, that is, prototypical pictures, of all four. Let us assume that the person in question identified a girl's foot. In this case the mental image will evoke an idea (interpretant in Peirce's terminology) of a girl who was there, who produced some walking activity in the direction indicated by the position of the toes in the footprint. The idea (or interpretant) will not be a concrete thought— it will be an idea that may involve all possible girls having that shape and size of foot. The idea would only turn into a concrete thought if the person had identified the footprint as belonging to, for instance, his daughter, if the person had established an iconic—and not an indexical—relation between the expression unit <footprint> and the pictorial content *his daughter's foot*. However, the idea itself will provoke other thoughts and give various associations. The thoughts and associations will have a ping pong effect on the mental image itself in the sense that it will gradually grow into a more and more concrete depiction. If in the course of the unlimited process of semiosis the observer suddenly realizes that it must be his neighbor's youngest daughter, the mental image finally turns into a picture and the idea to a thought. In that way the observer arrives at an objectively-founded explanation of the footprint in the mud. Despite that, it will still be a hypothesis, because the picture was self-produced and not a result of what he actually saw at that time the/a girl walked and left the/a footprint in the mud.

The reasons for re-examining Parker's footprint scenario should be obvious. First, I wanted to show that Peirce's (immediate) object is equivalent to a mental image which may be a visual, auditory, olfactory, gustatory or somatosensory one. At any rate, it cannot be the concrete object itself, that is the foot that made the footprint. The argument for stating that mental image seems to be an appropriate term is delivered by Peirce himself: the three sign types, namely icons, indexes and symbols, are established by Peirce himself by relating representamen to its (dynamical) object (cf. EP 2, 483-91; CP 4.536, note 1). It is difficult to realize that there could be an iconic relationship between a representamen and its (dynamical) object without involving the notion of image or picture. This is, I think, a firm piece of evidence for my claim—so firm that it seems difficult to ignore it. Secondly, I wanted to show that it is necessary to operate with a sharp distinction between what Hjelmslev called form and substance, or what Peirce called type and token. The distinction between form and substance with respect to object and interpretant is also made by Peirce himself, but not in his earlier writings. The distinction is made in his later writings, in his so-called third period which includes the important correspondence with Lady Welby (cf. Peirce, 1953, where his letters to Lady Welby are published). Here he differentiates immediate and dynamical object, on the one hand, corresponding to my distinction between the abstract image and the concrete picture of an object, and immediate and dynamical interpretant, on the other, corresponding to my distinction between the abstract idea and the concrete thought. He writes to Lady Welby, March 14, 1909:

suppose I awake in the morning before my wife, and that afterwards she wakes up and inquires, "What sort of a day is it?" This is a sign, whose Object, as expressed, is the weather at that time, but whose Dynamical Object is the impression which I have presumably derived from peeping between the window-curtains. Whose Interpretant, as expressed, is the quality of the weather, but whose Dynamical Interpretant, is my answering her question. ... The Dynamical Interpretant is the actual effect that it has upon me, the interpreter. But the significance of it, the Ultimate, or Final, Interpretant is her purpose in asking it, what effect its answer will have as to her plans for the ensuing day. (CP 8.314)

Peirce treats the immediate object as an imperfect/incomplete representation, but, nevertheless, it is the immediate object, not the dynamical one, that is represented in the sign. In contrast to that, he treats the dynamical object as the perfect representation, the object as it really is (see, for instance, CP 4.536). The former seems to be more or less equivalent to what I call a mental image, that is an abstract prototypical picture of a certain object which is part of a linguistic sign. The latter is the concrete reality, but because concrete objects cannot cause mental images by themselves, but only through concrete pictures and because concrete objects cannot be identified by human beings without having visual or other kinds of pictures of them, I will argue that dynamical object should be regarded as being equivalent to picture, that is a concrete visual manifestation of a specific object that is either being looked upon or which has been evoked from the past or present world stores of our memory. Peirce himself calls it impression (cf. above), but if we use this notion, it becomes really difficult to keep the dynamical interpretant apart from the dynamical object. The important thing is that picture (dynamical object) is not part of the linguistic sign, but will normally be associated with it when the sign is used in concrete discourse. Mental image is form, picture is substance. Ransdell (1977, p. 169) formulates this in the following way: "[the immediate object is] what we at any time suppose the object to be ... [it] may fail to include something that is true of the real object." For him the immediate object is how we understand the object at any point in the semiotic process, whereas the dynamical object is the actual object—how we understand it at the end of that process. The dynamical object is the goal that drives the semiotic process.

Although Peirce distinguishes three kinds of interpretants (cf above), I allow myself to exclude the so-called final interpretant (it is defined in many different ways by Peirce). If we do that, Peirce says:

In all cases [the Interpretant] includes feelings; for there must, at least, be a sense of comprehending the meaning of the sign. If it includes more than mere feeling, it must evoke some kind of effort. It may include something besides, which, for the present, may be vaguely called "thought". I term these three kinds of interpretants the "emotional", the "energetic", and the "logical" interpretants. (EP 2, 409)

I here agree with Short (2004, p. 235) who states that the immediate-dynamical distinction describes the interpretant at different stages in the semiotic process, while the emotional-energetic-logical trichotomy describes the types of interpretant which are possible at any stage of the semiosis process. Apart from the fact that Peirce includes emotions into the interpretant, whereas I put them in the immediate object corresponding to my mental image, it seems to be reasonable to assume that the immediate interpretant is close to my notion of idea and dynamical object to my notion of thought. The immediate interpretant is a mere possibility—nothing can guarantee that it will ever be realized. The dynamical interpretant is the actual effect in the interpreter, while the immediate interpretant incorporates what is common to different understandings of an object. Peirce also writes the following:

> In regard to the interpretant we have equally to distinguish, in the first place, the Immediate Interpretant, which is the interpretant as it is revealed in the right understanding of the Sign itself, and is ordinarily called the meaning of the sign; while in the second place, we have to take note of the Dynamical Interpretant which is the actual effect which the sign, as a Sign, really determines. (CP 4.536)

In the following subsection I shall illustrate the image and the idea aspects of an expression unit. Below I try in a schematic form to sum up what we have been talking about so far.[2] I emphasize that by using the notion of picture instead of the real object we get a certain harmony between the dynamical object and the dynamical interpretant. They are not part of the sign itself, but, nevertheless, they both play a significant role in communication. What the speaker sees and thinks about a specific object (token) determines his or her choice of sign and what the hearer sees and thinks when being confronted with that sign is not only determined by the image-idea aspects of the sign understood as a type, but also by the specific object itself (if it is known to the hearer) or by his or her previous acquaintances with other tokens of the same type (if it is not known to the hearer).

Table 2: Correlations Between Peirce, Hjelmslev and Present Work

Peirce	Present Work	Peirce / Hjelmslev
Dynamical Object	Picture	Token / Substance
Immediate Object	Image	Type / Form
Dynamical Interpretant	Thought	Token / Substance
Immediate Interpretant	Idea	Type / Form

2.4. Image-idea Pairs in Practice
  2.4.1. Arguments for the three-sided sign
The double nature of the content side of simple linguistic signs (i.e. lexemes) explains why and how people can relate entities in reality and in their mind to linguistic entities, and vice versa. When receiving a picture of a certain entity, for example,

---

2. I emphasize that the schema presented above is meant to be an illustration of Peirce's terms applied to human language. I do not pretend that these correlations hold good for all other sign systems. Language is a specific and complex sign system and as such it need not have any other equivalent.

book, any person not suffering from any kind of aphasia is capable of describing it (e.g., an object consisting of bound pages with words written on them or pictures drawn on them) and to apply the right name to it (e.g., *book*). Similarly, hearing a description of an object, any person will be capable of drawing a picture of it and at the same time name it. And last, but not least, hearing a word any person is capable of drawing a picture that matches its referent and give a description of it. It is not possible to account for all this by using the traditional Saussurean approach that lacks the image side. It is exactly the mental image that accounts for the link between reality itself which offers a lot of concrete pictures (corresponding to Peirce's dynamical objects) and the mind which produces a lot of concrete thoughts against the background of varieties of pictures (corresponding to Peirce's dynamical interpretants).

2.4.2. Explaining the New Sign Conception

Let us assume that some person has never seen a certain object denoted <X> and has no background knowledge of the object, X. Let us assume that X is shown to the person and he is told of everything that is worth knowing about X. I argue that on the basis of the received picture of X the person creates a mental image (corresponding to Peirce's immediate object) which is a prototypical picture (for experimental evidence, see, for instance, Piaget & Inhelder, 1966; Ritchey, 1980; Beaugrande, 1985; Krampen, 1986, 1990; Dodge & Lakoff, 2005; and Gerlach, 2008; for other arguments, see Sebeok & Danesi, 2000, p. 1ff). The image is assumed to be coupled to an idea which is an abstract and structured representation of what he or she has been told or otherwise learnt about the object (corresponding to Peirce's immediate interpretant). Let us, moreover, assume that X has been introduced to this person by the name *tennis court*. Being a common noun the name *tennis court*—although fixed to a particular object during dubbing (This is the term used by Devitt & Sterelny, 1987)—is not grounded in the specific picture of that object, but in the prototypical picture, that is, the mental image of X. Likewise, the name *tennis court* is not grounded in a concrete thought of X, but in an idea, that is, a prototypical description, of X. The noun or nominal lexeme *tennis court* creates an image and an idea of it, and the expression unit itself pairs or mediates its image and its idea.

What is normally called the concept of tennis court is neither identical to the image of it, nor to the idea of it—the concept of tennis court incorporates the image side as well as the ideational side of it. The concept of tennis court is thus an amalgamation of the image and the idea associated with it. Since *tennis court* names the concept, its meaning is, so to speak, a particular image-idea pair. The peculiar thing about human languages is the property of lexemes of having two functions in one: they seem to name a single piece because there is only one expression, but actually they create two different pieces of content, that is, the image content made up of a prototypical picture and the ideational content made up of a prototypical description of the picture. In that way a lexeme functions as what is called an *engram* in neuropsychology: it contains many different kinds of information linked to one another in one single unit. This way of defining a lexeme makes language a tool that mediates perception and cognition—without language human beings would have difficulty in separating one from the other, on the one hand, and comparing them, on the other.

2.4.3. Three Nominal Naming Strategies

The traditional two-sided sign conception leaves no room for describing or explaining the different lexicalization patterns observed in various languages, for instance in Danish, French and Chinese (cf. Baron, 2002; Herslund, 2000). The division of the linguistic sign into two pieces of content makes this possible. It appears that there are three different ways of naming: one may specify (1) the image, (2) the idea or (3) the image as well as the idea. A language has to make a semiotic choice by applying one of three different naming strategies, that of specifying the image or the idea, or both at the same time. *Tertium non datur*. The existence of these three naming strategies becomes clear when, for instance, one compares the lexicalization patterns of Danish with those of English and French, on the one hand, and that of Chinese, on the other. It strikes one's attention that English common nouns such as golf *course*, ice *rink*, tennis *court*, running *track*, bowling *alley*, soccer *field*, *range* (in the Army), and *lane* all are called *-bane* in Danish. This cannot be a mere coincidence. If we stipulate that *-bane* has the meaning "location for rule-guided motion," we can account for all of its uses. This is surely not a specification of the image, but of the idea—more specifically a specification of what could be called function. In other words, the Danish nouns *golfbane* (golf course), *løbebane* (running track), *tennisbane* (tennis court) specify the ideational side of the noun by describing the function, that is, *bane*, and the subfunction (e.g., *golf-*, *løbe-*, *tennis-*) of the artifact in question.

If we compare all what is called *–bane* in Danish with the corresponding nouns in French, it appears that they mainly fall into three groups: those having *terrain* (e.g., golf course or football field), those having *piste* (e.g., ice rink or running track) and those having *cours* (e.g., tennis court or badminton court). That is to say, although French people very well know that *terrain, piste* and *court* are united by being different locations for rule-guided motion, the French language has chosen another naming strategy than the Danish one (for a detailed description, see Baron, 2002). It appears that it makes good sense to say that French lexicalizes the image structure, that is, what from the point of view of the human eye looks similar in structure gets the same name. This may also explain why French systematically lacks words for collective concepts of so-called artifacts that have no original in reality—French has many words for different chairs, for different machines, for different bowls, etc., but no words for the collective concept of chair, machine, bowl, and so forth, simply because there are no images, or immediate objects, for them—it is the idea of a common function that unites various chairs, various machines, various bowls, and so forth. Chairs in a dining room look different from chairs in a living room which differ from chairs in an office which again do not look like chairs in a café, and so forth. But despite that, they serve the same function which is part of the idea of a chair. The distinction between image recognition and function recognition has not only been demonstrated by people suffering from various forms of aphasia (some people may describe an object, but may not recognize it when looking at a picture of it and some people may recognize an object on a pictorial basis, but are not capable of describing the function of it), but the distinction has also been verified by brain scanning human being's cerebral processing of so-called nature facts, for instance, animals and flowers/plants, and artifacts, for instance, chairs and jugs: Nature facts seem to have a collective concept in the shape of an image (for instance, an image of a human being, that is neither male nor female), whereas artifacts seem to have a collective concept in the shape of an idea where the function is described (cf. Gerlach, 2008). This might be explained by human beings' long co-existence with other living creatures and with their relatively short-term experience of artifacts.

In Chinese, a third pattern appears. Chinese is very different from languages spoken in Western countries. One of the specific features is that every name must mean something, including personal names and names for places and products. Let us take some illustrative examples. If we translate *David* and *Mary* into Chinese, we get the following: *Dàwèi* and *Mǎlì*. *Dàwèi* means big (=Dà) protector (=wèi), while *Mǎlì* means beautiful (=Mǎ) treasure (=lì). What we observe here is prototypical of Chinese. The name is composed of two syllables corresponding to two morphemes— this is the standard pattern of nouns in Chinese. It is so typical that nouns that are composed of one syllable corresponding to one morpheme are duplicated without change of meaning. This concerns, for instance, *Bàba* (father) and *Māma* (mother), where the first syllable/morpheme of each name means father and mother, respectively, but are duplicated in order to sound (or to have two characters) like a typical Chinese word. The explanation for this is simple. The first morpheme names the image side, that is, big and beautiful (the qualities appear immediately from the receiving picture), whereas the second morpheme names the idea side, that is, protector and treasure (they cannot be visualized, only understood as being parts of different categories). It thus seems as if Chinese literally repeats our definition of a sign, namely that an expression unit mediates an image and an idea. What has been said here concerns all nouns, but it does not concern grammatical expression units and particles which are prototypically monosyllabic and have no genuine naming function.

2.4.4. The Diversity of the Image Side of the Term – The Sign of Common Nouns
Unlike the ideational content of a lexeme which has only one mental representation (consisting, however, of several descriptors, namely Category, Function, Subfunction, Location, Means, Object, etc.), the image content is not restricted to the visual image mentioned and more or less implied above (see figure 1). A lexeme such as *rice pudding* will not only mediate an idea (involving descriptors such as Artifact, Food, Mush, Dish, etc.) and a visual image (giving, for instance, appearance, form, dimension and color), but certainly also an olfactory, a gustatory and perhaps even a somato-sensory image. Without stipulating these images it is impossible to explain how and why people are capable of recognizing rice pudding when smelling it, tasting

it and touching it without looking at it or even knowing that it is present. It is important to note that their existence plays a crucial role in associative processes. If you see rice pudding in a picture, this will not only provoke an idea of it, but certainly also a (moderate) taste or a (moderate) smell of it. The same effect can be observed when being confronted with the expression itself, for instance, if you are hungry. The existence of these association links is partly the reason why companies are willing to pay huge sums for commercials.

Figure 1: An Illustration of an Image-idea Pair of a Common Noun

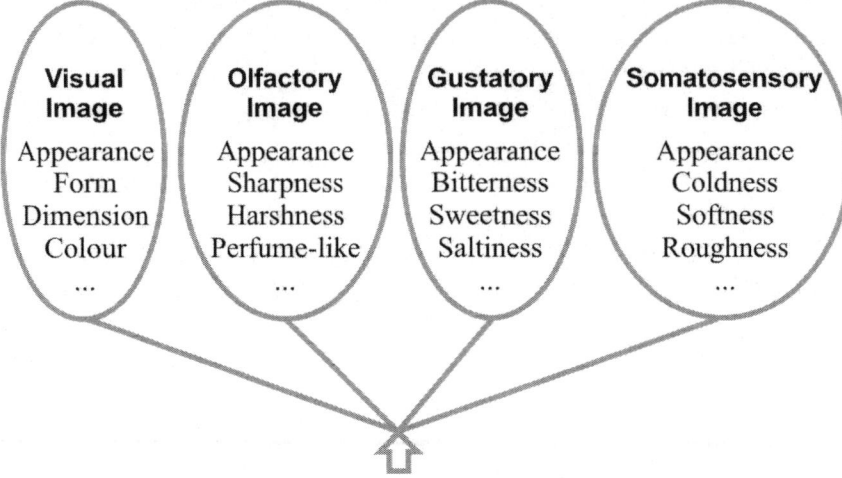

RICE PUDDING   [Auditory Image]

Idea
Category: Artefact
Subcategory: Food
Type: Mush
Function: Dish for Human b.
Ingredients: Rice grain/Milk
Procedure: Bla bla ...

2.4.5. The Strict Nature of the Image-idea Pair of the Rheme—The Sign of Verbs
As already mentioned above, action verbs such as *give* mediates two images, so-called ground-situations, and two ideas, so-called ground-propositions:

Figure 2: An Illustration of an Image-idea Pair of a Verb

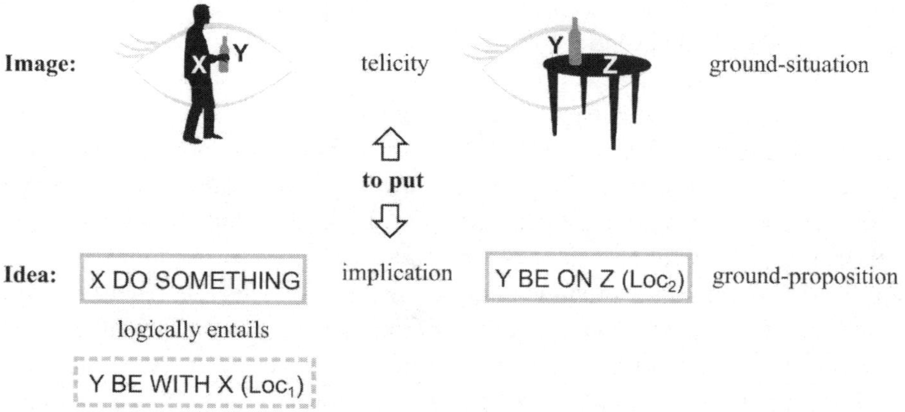

In the ground-situational structure there is an unstable image, an activity, in which X, the agent, is placing Y, the patient, on Z, the location-recipient, and a stable image, a state, in which Y is on Z. They are related to one another by the relation of telicity which says that they are tied together and that the direction goes from the activity towards the state. The ground-propositional structure describes what the unstable and the stable images depict. We have an activity description, X DO SMTH, which logically entails that Y BE WITH X, and a state description, Y BE ON Z, that is related to the activity description by the logical relation of implication. I call this a verb model, in this case a verb model of an action. All action verbs will create similar verb models, but some action verbs will not create visual images as, for example, *put*, but olfactory, gustatory, auditory or sensomotoric images.

It is interesting to note that Peirce seems to have had more or less the same kind of thoughts. Peirce elaborates on the example, *Anthony gave a ring to Cleopatra*, and says the following:

> Thus, "gave" conveys its meaning because the interpreter has had many experiences in which gifts were made; and a sort of composite photograph of them appears in his imagination. (CP 5.542)

This seems to be very close to my ground-situational structure which involves an activity as well as a state in the form of mental images. But Peirce seems also to be aware of the complex ground-propositional structure:

> The terminology of the older grammarians was better, who spoke of the subject nominative and the subject accusative. I do not know that they spoke of the subject dative; but in the proposition "Anthony gave a ring to Cleopatra," Cleopatra is as much a subject of what is meant and expressed as is the ring or Anthony. (CP 5.542)

In my ground-propositional structure *Anthony* (the nominative) is the underlying subject of the activity description, while *a ring* (the accusative) is the underlying subject of the state description. At other places where he mentions monadic, dyadic and triadic predicates (e.g., CP 1.288-292; CP 4.438; CP 5.469), he emphasizes that triadic predicates such as A sells C to B for the price of D can be analyzed as compounds of dyadic and triadic predicates.

2.5. Arbitrariness and Motivation of the Fundamental Simple Sign

There is no doubt that all lexemes, be they compounds, such as *golf course*, or non-compounds, such as *golf*, are symbols, that is, arbitrary and established by convention. But this is not tantamount to saying that lexemes are unmotivated. Their motivatedness does not appear in non-compounds (which does not imply that they are unmotivated), but only in compound ones. In other words, it is not a mere coincidence that golf course is named *golfbane* in Danish, but *terrain de golf* in French. And it is certainly not coincidental that *Mary* is translated into *Mǎlì* in Chinese. Nevertheless, the Danish, French and Chinese names are arbitrary—the relations between the expression units making up the words, on the one hand, and those image-based and idea-based elements making up their content are in no way transparent or predictable. They are arbitrary and must be learnt as conventions agreed upon by the speakers of the three different speech communities.

Except for Benveniste, who suggested replacing arbitrariness by necessity (cf. Benveniste, 1966, p. 47ff), nobody seems to have taken any specific interest in exploring the function of arbitrary lexemes. Jakobson (1962, 1963, 1965) criticized Saussure for considering language as such to be an arbitrary system, but all his examples of indexicality and iconicity are taken from grammar itself. Jakobson never really cracked the problem and actually avoided answering the question: For what reason are words arbitrary? What is the function of having a repertoire of pure symbols? To my knowledge, nobody ever asked these crucial questions. The same holds true of Jakobson despite the fact that he advocated a teleological approach to language and struggled all his life to develop a means-end model for language (cf. Jakobson, 1963).

2.6. The Function of the Arbitrariness of the Simple Sign

We shall keep to the sharp distinction between the lexicon of a language consisting of lexemes, which constitute an open class or an additive system, and its grammar consisting of grammemes which constitute a closed class or a structured system. Let us start by looking at the lexicon as an abstract entity. In principle, it contains names for three different things: 1) names for objects in the form of nouns; 2) names for situations in the form of verbs; and 3) names for qualities in the form of adjectives. Without grammar, no lexeme would be able to point out of itself—the common name *book* will create an image, but certainly not a specific picture of a particular book, and simultaneously an idea, but certainly not concrete thoughts about a certain book. Peirce uses the term *impotence* of common nouns, that is, a common noun is not capable of evincing reference by itself (CP 4.543). I agree, but prefer the term *static*, that is, all symbols are static and therefore need an index. But it would be strange, if arbitrariness only yields a negative quality. There must be a positive effect of making a name a symbol. The common noun *book* is impotent/static, but for what reason? For what purpose? The positive effect of making *book* impotent is its abstract capacity to refer to any book in the past, present and future, be it in a real or an imagined world. I call this the *omnipotence* of the word. All lexemes have this capacity simply because they are symbols where there is an arbitrary relationship between expression and content, be it the image-based content or the idea-based content which is mediated by the expression unit. In short, arbitrariness makes words impotent in order to make them omnipotent: *book* may refer to any book in my book shelves, but it cannot do it alone—it needs a grammatical vehicle in the form of an article (e.g., *a* book or *the* book) or a pronoun (e.g., *this* book or *that* book), that is it needs an index in order to be able to refer.

As argued for in detail in Durst-Andersen (2008), the word *morning* applies to any morning in the past, present and the future—there are no limitations at all as long as *morning* is used according to its semantic potential. In the same way, *Morning!* can be used as a greeting every morning and among all people in an English-speaking community. If we look at the corresponding non-verbal greetings such as kissing, embracing, hand shaking, nodding one's head and waving one's hand (all functioning as so-called emblems or intentional gestures, cf. Kendon, 1995; McNeill, 1992), none of them would be impotent and none of them would possess the same universal character or omnipotence as the verbal greeting, because—and this is crucial—they all inherently contain situation-bound indexical and iconic elements, as I will try to demonstrate in the following.

When produced in a non-verbal greeting situation, all the symbol-like gestures can be said to involve three types of content (cf. Jensen, 1999). First, they all involve "I hereby show you my respect"—this is shown by the gesture itself and its direction. Secondly, they all involve "I hereby want you to do the same as I did." Thirdly, they all involve "We hereby reconfirm our mutual relationship" which is the effect of the addressee's reaction to the sender's message, that is, the effect of a successful communication. The point is, however, that although I can say *Morning!* to my wife, to my dear friend, to my colleague, to a person I pass everyday in the subway and to a total stranger whom I see for the first time in the front of my office, I cannot use any of the non-verbal greetings in all the above-mentioned situations. I cannot—or should I say: I will not—give a good-morning kiss to a person who is a complete stranger, or a good-morning nod with my head to my wife, or give a good-morning hand shake to the person I pass every day in the subway.

Each of the non-verbal greetings is limited in use, and they are so because as signs they are non-arbitrary: there is a diagrammatic, that is, an iconic relationship between the signifier and the signified: the physical distance between them reflects the intimacy or the depth of the mutual relationship between them. This means that if their relationship is intimate or deep, they will, if possible, use a gesture with no distance; if their relationship is non-intimate or non-deep, they will use a gesture with some distance. The point is that because of that, none of the gestures can fulfil the universal function of the verbal greeting. The fact that they involve iconic elements makes them speak with a concrete voice, and the fact that they are indexical, make them goal-directed and dynamic. In the case of gestures functioning as symbols it is not possible to separate the grammar from the actual signs used. The grammar is an inherent part of the signs themselves, because the signs are not only symbolic, but inherently indexical and iconic. In short, because they are not a hundred percent arbitrary, they cannot be omnipotent. What has been said at this place is more or less true of all non-verbal communication systems.

Before concluding, I emphasize that I have been talking about English-speaking communities with their specific codes for verbal and non-verbal greetings. Rules for verbal greetings may vary considerably depending on speech community. In some speech communities the genetic, the social and the situational role distances between the speaker and the hearer are grammatically encoded (this concerns, e.g., the Japanese and the Korean speech communities) and in others the presence of an audience may change intimacy into non-intimacy—grammatically speaking, of course (this concerns, e.g., the Russian speech community). Rules for non-verbal greetings may vary in the same way. In some communities one may not be allowed to greet a stranger (as in Brazil) or a male may not be allowed to shake hands with a female (as in Russia). But the conclusion remains the same: The grammar of non-verbal communication systems is an inherent and non-separable part in each non-verbal sign, whereas the grammar of verbal communication systems is an external and separable part. And this is not a coincidence.

2.7. Concluding Remarks

The arbitrariness of the linguistic sign is therefore essential for any communication tool to be used globally. Martinet wrote an important paper in 1949, in which he argues that language consists of two articulations—a distinction that is generally accepted, but which has never been elaborated upon. According to Martinet, the first articulation system is made up of morphemes or monemes as he calls them, that is, minimal signs, which together form words that can be combined into sentences. The second articulation system is made up of expression units, that is, phonemes, that do not in themselves mean anything, but whose function is to distinguish one sign from any other sign. According to Martinet, it is exactly the presence of the second articulation system that makes the linguistic sign arbitrary and, moreover, guarantees the economy of language. Although these qualities may sound trivial, they are nevertheless highly important. My point is, however, that there are some costs involved when one creates an economic system of arbitrary signs, that is, symbols. The consequence is that any linguistic symbol is purely static in itself and therefore completely helpless (in Peircean terms: impotent) in a communication situation that is dynamic per se. In order to be omnipotent it must be arbitrary, that is, static. A symbol therefore needs a vehicle, that is, a grammar, which can bring it to the proper place. A symbol names, but it cannot by itself refer. A symbol needs an index as well as an icon in order to refer. This is exactly what grammar provides. *Man* names, but *a man* or *the man* both refer—in the former case to the speaker's experience, and the latter case to the hearer's experience. *Man* is static, *a man* and *the man* are dynamic—they reach their semiotic goals because of the attached grammemes. I shall return to this important issue below.

3. The Complex Sign as a Dynamic Unit – An Index

When speaking of language as a communication tool we leave the level of lexemes and enter the level of utterance, that is, the substance behind the form *sentence*. At the same time we leave the level of simple signs and enter the level of complex signs, or supersigns in Eco's terms (cf. Eco, 1975). As mentioned in the introduction, nobody in contemporary linguistics has paid attention to this level from a purely semiotic point of view. Presumably, there are many reasons for that, but one of the most obvious ones is that all scholars have been taking Bühler's *Organonmodell* more or less for granted—even those who do not credit him directly (e.g., Searle whose distinction between assertives, expressives and commissives was unthinkable without Bühler's contribution). At any rate, the organon model has been the basis of any theory of language and communication developed after 1934, when the model was published in its final form (see Bühler, 1934). Nobody has ever questioned the model. It concerns the three obligatory participants of a communication situation and the corresponding three language functions, namely, the representative, the expressive and the appeal functions, and it also concerns its three semiotic correlates, namely, the symbol, the symptom and the signal. In the following, I shall attempt to show that there are hitherto unnoticed problems with his model. The problems are so fundamental that the model has to be revised and, moreover, it has to be supplemented with a model for the speaker. Bühler's *Organonmodell* is the hearer's model—he was interested in decoding and left out the encoding process, that is, the speaker's function in communication. Only by including the speaker, it is possible to reach the genuine level of the dialogue described by Bakhtin (cf. 1929/1994). Voloshinov is quite explicit in stating that "all utterances have an inherently dialogic character" (Voloshinov, 1929/1986, p. 25). Peirce was even more radical in arguing that all thinking is dialogic in form (CP 6.338).

3.1. Bühler's Language Functions Revisited
According to Bühler, the representative function is by far the most dominating function of language (cf. Bühler, 1934, p. 30), although each function is present in any utterance and therefore is part of the linguistic form, that is, Saussure's *langue*, and not only its substance, that is, Saussure's *parole* (cf. Hjelmselv, 1943): when a speaker utters "That horse is beautiful," the utterance does not only represent a certain state of affairs or a certain situation, but it also expresses the speaker's emotions (or judgement) and at the same time serves as an appeal to the hearer to give a response (see fig. 3). This is exactly what distinguishes Bühler's three language functions, namely the representative, the expressive and the appeal functions, from Jakobson's six (cf. Jakobson, 1960) and Halliday's seven functions (cf. Halliday, 1975). Bühler's

three functions are present in any utterance, whereas, for instance, Jakobson's poetic function may be present or may not be present (cf. Jakobson, 1960). Bühler's functions are language functions, Jakobson's and Halliday's are a combination of language and speech functions. As I see it, both mixed form and substance instead of keeping them strictly apart.

Figure 3: Bühler's Organonmodell

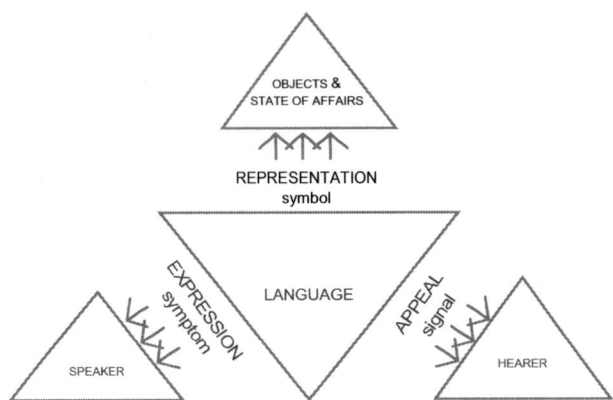

Although all this seems reasonable and convincing, a problem arises, when we read Bühler more carefully. While his expressive function and appeal function only make sense when language is used in communication, his representative function (in German *Darstellungsfunktion*, cf. the subtitle of his monograph) seems to be used in two very different ways corresponding to his two terms *objects* and *states of affairs* (Bühler, 1934, p. 30). Language represents objects, but it also represents states of affairs, that is, situations. However, there is a considerable difference between these two kinds of representation. When language represents objects, it need not do this in communication—quite to the contrary. In principle, this function is purely static and therefore has a place outside communication. It is reflected in dictionaries in the form of lexemes or words which make up the lexicon of a language. This is the static meaning of the terms *representation* and *Darstellung*. When language represents states of affairs, that is, situations in a real or an imagined world, we do not deal with words, that is, symbols, but with utterances or complex signs in Bühler's terminology. In short, we deal with language in use, with written or spoken discourse that is a dynamic, and not a static phenomenon. This constitutes the dynamic meaning of the terms *representation* and *Darstellung*. This duality is not found in the two other language functions, only in the representative function. And it is no coincidence: Bühler explicitly states (Bühler, p. 30) that language is a two class system of representation means, one for words, that is, the lexicon of a language, and another for sentence building, that is, syntax (see also pp. 69-78 where he elaborates on the distinction between words and sentences). According to Bühler, both systems perform a representative function, and it is exactly the double nature of the representative function that makes it the dominant function of language and distinguishes it from the other two functions: "in order to get it right at the place in the figure [figure found in Bühler, p. 25; corresponding to the non-revised model] where you find 'things' [Dinge] we now write the double designation 'Objects and States of Affairs' [Gegenstände und Sachverhalte]" (Bühler, p. 30). It looks as if Bühler was trapped by the German language which has one dynamic word, that is, *Darstellung*, for two different functions, the one being static and the other being dynamic. The same linguistic trap is found in English, where the static word *representation* is used for both functions.

I conclude the discussion by arguing that we have to keep simple signs separate from complex signs, that is, the naming function from the communicative function. In order to make sense from Bühler's organon model we must remove the word and its naming function from the model. This leaves us with the representative function used unambiguously and in line with the expressive function and the appeal function. All three functions are only present in communication and they together form the communicative function of human language. The three functions are obligatory manifestations of the communicative function, and, due to the implementation of grammar, they are present in any situation of communication. Words are symbols— are static and cannot by themselves take part in communication. We saw this with the example of saying *Morning!* in which the intonation revealed that it was an utterance, that is, a complex sign intended as a greeting and not merely a word, that is, a simple sign that was intended as a name for a particular part of the day.

3.2. Bühler's Sign Types Revisited

Bühler's organon model also comprises a semiotic elaboration of his language functions (see figure 3). According to Bühler, any linguistic sign, for example, a word, or any complex linguistic sign, that is, an utterance,

> is a *symbol* in virtue of its assignment to objects and states of affairs, a *symptom* (evidence, indication) due to its dependence upon the speaker whose innerness it expresses, and signal because of its appeal to the hearer whose external and internal behaviour it governs just like other communicative signs (Bühler, 1934, p. 28).

Hence, the three language functions become explicitly linked to three different sign types. Thus Bühler offers a semiotic model of linguistic signs which Peirce never did. Interestingly enough, nobody ever questioned the semiotic triad, presumably because each of the three sign types makes sense if one pairs them with their respective language function and ignores the duality of the representative function. The problem arises when we look at the triad itself, that is, symbols, symptoms and signals. Symptoms and signals are clearly indexes. A symptom points backwards to the speaker and therefore we also go back in time, that is, back to an earlier experience. A signal points to the hearer and therefore points forward in time. A symbol does not point, but names. There is no motion or direction involved, at all. And time is not involved, either. A symbol is achronic, that is, timeless. In short, the triad consists of unequal elements. There seems to be a mismatch.

Again we suspect that the mismatch is due to Bühler's lack of distinction between words as static elements and utterances as dynamic elements. It makes sense to claim that words are symbols of objects. But it does not make sense to say that a specific utterance is a symbol of a certain state of affairs or a certain situation—it makes only sense if an utterance is transformed into a string of logical symbols, but in that way the logical notation will be a symbol of the utterance, not of the situation denoted by the utterance. If all utterances were complex symbols consisting of simple symbols, it would follow that we had to distinguish between two sets of symbols, that is, static and dynamic ones. It also follows—and this is really important—that there would be an arbitrary relationship, established by convention, between every single utterance and every single situation. This would mean that all persons would have to learn all utterances by heart in order to understand what they symbolize and in order to produce them later on—just as they learnt all words by heart in order to understand the meaning of them and later to be able to reproduce them in utterances. This completely contradicts common sense and completely undermines what is normally referred to as *the productivity of language*: due to its grammar language can give expression to any old and any new thought, and they can all be understood by the hearer, if he or she masters the language in question (see, for instance, Hockett, 1963; Jakobson, 1968; & Lyons, 1977, p. 76).

In short, words are symbols, but utterances cannot be (which is not tantamount to saying that they cannot be used symbolically). Utterances must function as a sort of index, that is, perform a pointing function which makes them dynamic and goal-directed at the same time. The consequence is that we have to remove *symbol* from Bühler's triad and instead look for a lacking index. We are looking for a concept for the sign of that part of reality that was stimulus to the speaker's experience, of which the sign is also symptom and to which the hearer returns when he has mentally processed the effect of the same sign understood as a signal. Or, to put it more simply: we are looking for a sign that points to the situation that is stimulus to the speaker's experience and at which the hearer arrives after having decoded the utterance. I shall call the lacking index *model*, and define it as the semiotic sign correlate to the representative function of language: an utterance is a model, because it points to a situation.

This means that we get the following indexical triad: model, symptom and signal. The triad corresponds to another triad, namely situation, the speaker's experience of that situation, and information to the hearer concerning that situation. The order is not random: A piece of information to the hearer presupposes an experience by the speaker, and the speaker's experience presupposes a situation. Without a situation, there is no experience, and without experience, there is no information. In the same way, model, symptom and signal constitute a corresponding order within the realm of indexicality: to use Peirce's terminology, *model* is Firstness, *symptom* is Secondness, and *signal* is Thirdness. According to Peirce, indexes themselves belong to Secondness, whereas icons belong to Firstness and symbols to Thirdness. This means—still according to Peirce—that a symbol is based upon an index and an index upon an icon. We shall return to that later.

3.3. The Revised Model – The Semiotic Wheel

If what has been stated so far is true, we can say that Bühler's organon model with the three language functions is a *decoding model* for the hearer. The model explains why the hearer has to interpret any utterance in three different directions in order to draw all pieces of information out of it. An utterance will always point at a certain situation, express the speaker's experience of it and appeal to the hearer to find out the information status of the utterance alone as well as of its parts and through mental models get access to the specific situation referred to. In other words, the organon model is *the hearer's model*, in which we are dealing with a both-and relationship, not an either-or relationship.

The hearer knows that any utterance in any language involves a model of a situation in a real world or in an imagined world, a symptom of the speaker's experience of that situation and a signal to the hearer to find a match in his memory and via mental models get access to the situation itself. This was what Bühler wanted to state in his organon model, but although we revised it by removing *objects* and *symbol*, we did not succeed in changing its purely static look at the expense of a dynamic one. In order to do so we have to transform the triangle into a wheel (see fig. 4).

Figure 4: Fig. 4: The Hearer's Model – The Semiotic Wheel

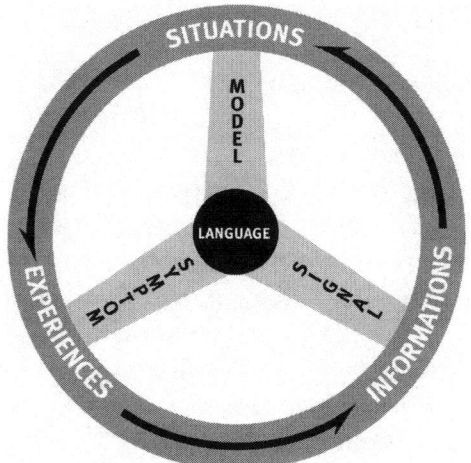

The Semiotic Wheel is constructed in such a way that the hearer must take a full tour of the semiotic wheel, irrespective of the place in which he or she arrives. My point is that different languages arrive at different places because of their different grammars. The Russian hearer arrives at *Situations*, but has to make the entire tour via

Experiences and *Informations*. The Bulgarian hearer arrives at *Experiences* and has to make the rest of the tour on his own, that is, via *Informations* and via *Situations* back to *Experiences*, but now in a completely different position, because he or she has drawn all possible pieces of information out of the utterance in question. The English hearer arrives at *Informations* and has to take the rest of the tour on his own, from *Informations* via *Situations* and via *Experiences* back to *Informations* in order to make the final interpretation of the utterance, in which the different parts are put together (see also subsection 4.2 and section 5 below; for a more detailed account: see Durst-Andersen 2005b, 2008a).

In short, any utterance in any language will serve for the hearer as a model of a situation, as a symptom of the speaker's experience of the situation and as a signal to find a matching experience and a matching situation to it. Since everybody is aware of this, they must know it not only in their capacity as hearers, but also in their capacity as speakers.

3.4. The Natural Conclusion
The natural question arises: which sign and communication model functions as the point of departure for the speaker? Which model will guide the speaker when he gets an intention and wants to verbalize his intention? We are talking about an *encoding model*, in which we have to deal with the crucial question "What is the function of grammar in communication?," or "For what reason does language need a grammar?". Bühler did not ask such questions. We know that the main function of language is to serve as a communication tool of human beings and that it fits this function and all subfunctions derived from it. But nobody addressed this question with respect to grammar. And nobody ever considered its function in a communication situation. We will try to do that in the next section.

4. The Semiotic Function of Grammar

4.1. From simple signs to complex signs
The answer to the question, "What is the function of grammar in communication?" is quite simple: since all lexical units of a language have an arbitrary and completely conventional relation between expression and content, language must have a grammar that can instruct the hearer how an expression should be understood. The omnipotence of each symbol, which is ensured by its arbitrariness, would be empty without a grammar, because grammar is the tool or vehicle that carries the name to the right place. Precisely because of its arbitrariness, language must have a grammar that can give a *semiotic direction* to the otherwise completely static sign or lend a *voice* to the otherwise completely mute sign—I here use the important Bakhtinian notion of voice. The fact that lexemes are symbols and thus arbitrary yield them no semiotic direction at all. They are all static, and the *staticness* is their great advantage and at the same time their weakness. Their static nature makes them prepared for any kind of job, but at the same time they are totally dependent on a vehicle in order to get a specific job

done. If there is any direction at all in a symbol, it is inner and not outer reference: A name will just refer to its own place in the linguistic system and at the same time create a combined image-idea in the mind of a human being, (cf. section 2). Symbols in themselves do not refer, they just name. In order for a noun, or a verb, or an adjective to get access to a specific object, or a specific situation with specific participants, or a specific quality they must be supplied with a device that draws the hearer's attention to what the speaker intends—and this is exactly what grammar (or any other indexical device) does. Another way of putting it is to say that *morning be good* (or in any other order) gives only a vague impression of what really is at stake— it has a referential potential, nothing else. In order for the speaker and the hearer to get access to a specific real or imagined world one has to use grammemes for situation type, time world and ontological status of the world referred to.

4.2. Grammar as a Prime Index

Grammar in itself functions as a *prime index* that makes symbols dynamic by giving them a semiotic direction. Due to the fact that a direction always contains a target, grammar can also be said to provide the symbols with a specific frame of reference. But unlike decoding which is a search in all possible directions, encoding involves an obligatory choice between three possible targets. Why? Because all communication requires three participants, namely reality, speaker and hearer, as Bühler told us. But what he (and Peirce) did not tell us is even more important. In fact, a *state of affairs* has three modalities of existence: (1) the situation as such as it is shared by the speaker and the hearer; (2) the situation as it is experienced or not experienced by the speaker; and (3) the situation as it is experienced or not experienced by the hearer (see also Durst-Andersen 1992 for a cognitive explanation of this three-way ambiguity). This means that one may refer to the situation itself, to the speaker's experience of it, or to the information intended for the hearer, which is the linguistic result of the speaker's comparison of his own experience with the hearer's—if they match, it is old information, if they do not, it is new information. The members of a speech community must agree on either of the three modalities of existence. In short, they must make a choice in order not to complicate matters for the speaker in his encoding process and not to confuse the hearer in his decoding process. The encoding and decoding processes would be extremely difficult if those parts that make up a grammar were completely disharmonic by pointing in different directions. We are thus not dealing with a both-and relation as we were in the case of decoding, but with an either-or relationship in the encoding case.

Before encoding the speaker has to make two choices that only in principle can be separated: 1) what do I want to say?, and 2) how should I put it? The former choice is entirely dependent on the speaker himself, whereas the latter choice is made by all members of the speech community to which the speaker belongs. This choice is a choice of type of grammar or code, or, as I shall call it, a choice of *linguistic supertype*. All languages have different systems, but these systems tend to group into three supertypes. In the former case, we are dealing with specific rules, in the latter case, with different principles. In other words, languages always differ with respect to rules, because a principle can be interpreted in various ways (cf. Coseriu, 1965/1975).

Figure 5: The Speaker's Model – The Grammatical Triangle

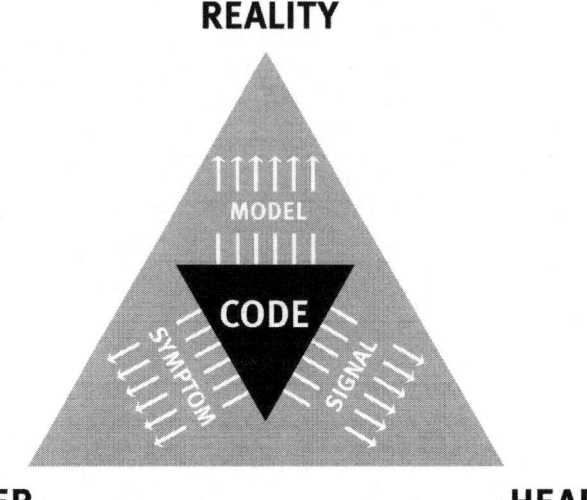

The speaker knows the potential three-way ambiguity of any symbol and in order to be able to provide the hearer with an unambiguous tool that can instruct him how the string of symbols should be understood, there is an obligatory choice between the three types of indexes established above, namely model, symptom and signal (see fig. 5). This means that all languages have a set of symbols, that is, a lexicon, but must choose between three types of indexes, that is, between three grammatical supertypes, the function of which is to be able to bring the symbols to their target by giving them a semiotic direction, that is, by pointing to situations in reality, to the speaker or to the hearer. Any symbol is omnipotent and is ready to be brought to one of the three targets. But just as a vehicle cannot drive into three different directions at the same time, a grammar cannot point to the situation, the speaker and the hearer at the same time. A language may talk either about situations, or about experiences, or about informations (for a more detailed account of the grammatical triangle, see Durst-Andersen, 2005a, 2008a)

5. Grammemes as Indexes and Icons

The semiotic direction is determined by the grammatical orientation of the supertype in question and is therefore outside the speaker's control. But on the basis of the semiotic direction indicated by the grammemes of a given language, the speaker must decide whether the target being pointed at is hit or not. The question is posed to any speaker and should be answered whenever he or she is producing an utterance. The answer is always of the same type, but due to the fact that there are three different targets, its consequences may be different.

1. Languages with a grammatical orientation towards situations shared by the speaker and the hearer pose, for instance, the question: Is the object present or absent in the situation referred to? In Russian the nominative, the accusative and the genitive of a noun all point to a specific situation—this is the integrative function of these three central cases determined by supertype. But whereas the nominative and the accusative say that the object is present in the situation referred to, the genitive says that the object is absent in the situation referred to (whereby it automatically refers to the speaker's and the hearer's world of common background knowledge)—this is their differentiating function.

2. Languages with a grammatical orientation towards the speaker's experiences of situations pose, for instance, the question: Is the object part of the speaker's experiences or is it not? In Bulgarian the so-called definite article and zero-form of a noun both point to the speaker's world of experiences—this is their integrative function determined by supertype. But whereas the definite article says that the object referred to is part of the speaker's world of experiences, the zero-form says that it is not part of it (whereby the form automatically refers to the object in a specific situation)—this is their differentiating function.

3. Languages with a grammatical orientation towards information to the hearer pose, for instance, the question: Is the object part of the hearer's experiences or is it not? In English the definite and indefinite article both point to the hearer's world of experience—this is their integrative function determined by supertype. But whereas the definite article says that the hearer has an experience of the object experienced by the speaker, the indefinite article says that the hearer has no experience of it (whereby it automatically refers to the speaker's world of experiences)—this is their differentiating function (for further information, see Durst-Andersen, 2005b, 2008a).

As it should appear very clearly, the choice made by the specific grammemes in Russian, Bulgarian and English is fundamentally based on iconicity: (1) does the copy named by the Russian noun (source) have an original in the situation referred to (goal)?; (2) does the original in the situation referred to (source) have a copy in the speaker's store of experiences (goal)?; (3) does the speaker's copy in his store of experiences (source) have an equivalent in the hearer's store of experiences (goal)? In other words, what we call case in Russian, articles in Bulgarian and articles in English all point in a semiotic direction but differ with respect to the source and the goal. Type of source and type of goal fit together. In the same way the members of grammemes sharing this indexical function will distinguish themselves from one another by marking plus or minus to the actual iconic relationship between the source of the direction and its goal. If there is no iconic relation, the three grammemes sharing this property necessarily talk about the source—the Russian genitive will talk about an imagined book, the Bulgarian indefinite article about a certain book placed in a specific situation, and the English indefinite article about the speaker's experience of a certain book. If there is an iconic relation, the three grammemes having this property necessarily talk about the goal—the Russian nominative and the accusative will talk about a certain book placed in a specific situation, the Bulgarian definite article about the speaker's experience of a certain book, and the English definite article about the hearer's experience of a certain book that is also experienced by the speaker.

As should be evident, there is a partial overlap between the three types of languages—in that way any language may take a further step and change its type. I emphasize that the fact that grammemes function as indexes and icons at the same time explains why Jakobson (and his pupils, cf. Shapiro, 1983) found iconic relations within the grammatical systems of various languages and why he constantly saw binary relations, that is, plus or minus (cf. Jakobson, 1962, 1963, 1965).

6. Discussing Icon, Index and Symbol

According to Peirce, there are three sign types when the expression side of a sign is related to its dynamical object, that is, our picture. It is crucial to underline that Peirce explicitly says that he is talking about its relation to the dynamical object, corresponding to a concrete picture, and not to its immediate object, corresponding to an abstract image (cf. Peirce, 1953: 1908 letter to Lady Welby). However, if one compares the contents of what was examined just above with Peirce's triad, namely icon, index and symbol, one may be confused. Peirce considered icons to be firstness, indexes to be secondness and symbols to be thirdness. This means that symbols presuppose indexes and indexes presuppose icons. According to Peirce, any symbol is built upon an index and this index is built upon an icon. But just above I said that any linguistic symbol needs an index that necessarily needs an icon. It is extremely important to keep the two approaches apart from one another. In the former case, we are dealing with *morning,* that is, the lexeme itself, in the latter case with nominal phrases such as *a morning* and *the morning*, that is, the word as part of an utterance in which we find grammatical affixes attached to lexemes. I shall argue that the inherent properties of a symbol, that is, that it incorporates an index as well as an icon, are repeated as external functions when it functions in an utterance that is to be decoded by the hearer. Let me dwell a little bit on this crucial point.

6.1. The Inherent Properties of a Symbol

We shall return to one of the possible morning gestures, that is, the good morning kiss. Let us assume that I kiss my wife on the mouth every morning. The very choice of the kiss is conditioned by our mutual relationship: we are as close as husbands and wives can get. Kissing one another on the mouth with no distance at all between the two persons involved is thus an icon of the psychological closeness between them. At the same time, my opening act, that is, my part of the kiss, is a model which lacks its mirror image, that is, my wife's part of the kiss, which closes it. Moreover, it is a symptom of my feelings and a signal to my wife to do something in return on the basis of the showed model. It becomes a symbol of our mutual love at the moment our mouths touch each other, but also a sign act in the sense that in so doing we reconfirm our mutual relationship—it is a kind of loving contract that is re-signed whenever we kiss one another. Communicating with one another by means of verbal or non-verbal signs is extremely important, because people resign their contracts with other people or with the society. The social function of language consists in signing and re-signing a contract which makes its signers legitimate members of a given speech community.

The point is that any symbol has to possess the inherent properties of an index and an icon. This also applies to the common noun *kiss*. It is a symbol, because the relation between its expression and its image/idea sides is arbitrary and conventional, but due to the fact that the word is learnt by the child in a concrete situation, where, for instance, the mother is kissing the child, the word will always be accompanied by the iconic meaning of no distance and the various indexical meanings. If the noun *kiss* did not possess these qualities, no person would be able to use it correctly, that is, according to the laws established by the members of the English-speaking community. *Kiss* was learnt in a concrete situation and therefore linked to it with all possible iconic and indexical ties, but later, the child understands that *kiss* is a symbol that can be used in any situation that resembles the first. It is a kind of extrapolation. At that moment, the child realizes that symbols are omnipotent, that is, the image content attached to it is a prototypical picture with no concrete faces or bodies involved and the idea attached to it is not composed of concrete thoughts of how it is produced, because the varieties are many. At that time, the child realizes that thirdness covers something general that can be repeated at any moment of time as something concrete or, as Peirce would put it, as a replica of the symbol.

6.2. The Inherent Properties as External Functions

When we attach the grammemes *the* or *a* to the noun *kiss*, we are talking about a specific kiss in a concrete situation. If I say to my wife: "A kiss from you would make me happy," I take my starting point in an experience of my wife's kissing me in an imagined world (that is to say it is only an imagination) and by using the indefinite article I show her that I am talking about a kiss that she has no experience of. In other words, my symbol *kiss* is actualized in my mind at a particular place, and the particular place forms the starting point of its semiotic direction towards its goal, a potential place in the mind of my wife. The goal is, however, not achieved, because there is no such experience in her mind.

If I say to my wife: "The kiss from you made me happy," I take my starting point in an experience of my wife's kissing me in a real world (that is to say I have a real experience), and by using the definite article I show her that I am talking about a kiss that she has an experience of, too. In other words, my symbol *kiss* is actualized in my mind at a particular place, and the particular place forms the starting point of its semiotic direction towards its goal, a specific place in the mind of my wife. The goal is achieved, because there is a similar experience in her mind—a *cominterpretant* in Peirce's terminology (cf. EP 2, 478).

The indexical and iconic roles of grammar being necessary in order for language to function as a means of communication, is not identical to the indexical and iconic properties of any symbol. This also appears from the fact that something is added to the noun, that is, either the definite or the indefinite article. But the fact that any symbol needs a vehicle (an indexical sign) to point to the semiotic direction taken by the symbol and the fact that any symbol being carried from a starting point in a certain direction need an indication of whether the goal is reached or not (an iconic sign or its negation) is no coincidence. I see this as a kind of repetition: The inherent properties of a symbol are repeated as external functions, but in the order of a mirror image:

Figure 6: Inherent Properties and external functions of an Actualized Symbol

$$[[(\text{Icon (Index (Symbol)))) Index] Icon}]$$

This is so because a symbol is an extrapolation of something concrete or an abstraction whereby the symbol loses all its ties to concrete objects or situation-bounded elements. In order to become omnipotent nothing in its expression must be bound to or completely mirror its original basis. The omnipotence of any symbol makes it, however, a powerful communication tool. In order to function in communication the symbol has to be taken out of its isolated and abstract world and be brought back to real life. This restoration process is taken care of by grammar.

In short, a child learns the meaning of a name on the basis of iconic and indexical elements, but the properties are extrapolated from a concrete world to any possible world by which the name becomes a symbol. These properties are repeated as external functions in order to make the static and mute symbol dynamic, that is, goal-directed and speakable, and in order to make it reach its goal, be it the situation, the speaker or the hearer. All this was not anticipated by Peirce himself, but it in no way contradicts what he says.

7. Grammatical Categories in Collaboration – Argumentative Signs

Until now, we have been examining Peirce's representamen and its relation to itself, that is, as a qualisign, a sinsign or a legisign (Firstness) and its relation to its (dynamical) object, that is, as an icon, an index or a symbol (Secondness). We now turn to its relation to the interpretant (Thirdness). At this level, Peirce distinguishes the rheme (first order), the dicisign or dicent (second order) and the argument (third order) (see table 3).

Table 3: Peirce's Basic Sign Categories

	Firstness	Secondness	Thirdness
Firstness **A sign (represent-tamen) relates to itself**	QUALISIGN	SINSIGN	LEGISIGN
Secondness **A sign relates to its object**	ICON	INDEX	SYMBOL
Thirdness **A sign relates to its interpretant**	RHEME	DICENT	ARGUMENT

Note: Borrowed in part from Sheriff (1994, p. 41)

Evidently, when speaking of human language, argumentative signs are traditionally associated with texts in which a certain combination of utterances gives a specific argumentative direction and structure. However, I shall try to demonstrate that deductive, abductive and inductive arguments found in various texts are externalized functions of inherent properties of the verb itself, that is, with all its grammatical and lexical categories. It appears that verbal categories communicate or collaborate with one another and the result of that collaboration is a logical or rhetorical argumentation. In other words, what we saw just above repeats itself here: What is found as inherent properties within the symbol itself, that is, iconic and indexical elements, has an external manifestation in the shape of grammemes the functions of which are to point to a target (the indexical function) on the one hand and to indicate whether or not the target is reached (the iconic function) on the other.

But my task is now exactly the opposite. Above, I tried to show the external functions on the basis of known inherent properties of a symbol. Now I start at the opposite end. We know that texts involve deduction, abduction and induction—this applies to oral as well as written texts. In other words, we know the external functions, but we do not know where they come from. I want to demonstrate that they are found in the verb itself as a result of the various kinds of interplay between different verbal categories, be they lexical-grammatical or purely grammatical. If this is true, communication is not only related to oral or written texts—it is an inherent property of the verb itself: the speaker involves the hearer by arguing deductively, abductively and inductively for a certain view on reality. In that way, it is possible, at this level, to incorporate the Bakhtinian notion of *dialogism* and demonstrate the dramatic effect of putting lexical and grammatical forms together. What we shall witness is an internal dialogue within the verb.

7.1. The Verb as Rheme, Dicent and Argument

According to Peirce (cf. Nöth, 2002), any verb in its infinitive form will be a so-called rhematic sign: it names a potential action with potential participants. This also concerns the French action verb *entrer* (to enter) which is paraphrased in the following way in my theory (cf. Durst-Andersen, 1992, 2008b): "X do something & X exist in L," where X is any human being, *do* stands for any possible activity p and "exist in L" for any possible existential position q inside a certain container or location, L.

All possible realizations of *entrer* in a grammatical, acceptable and appropriate utterance will involve a finite form of the verb, for instance the following past tense forms: *entra* (entered), *est entré* (has entered), *entrait* (was entering) (see below). Any finite form of the verb will necessarily denote either an *event*, that is, p caused q, or a *process*, that is, p intended to cause q. A past event can be denoted either by the *passé simple form*, *entra*, or by the *passé composé form*, *est entré*. Both forms will refer, that is, point to the same event and will therefore involve the same statement, that is, "X produced an activity which was sufficient for the fact that X exists inside L." This statement is composed of three assertions, namely, X produced an activity; the activity was sufficient; and X existed inside L at the time of reference. A past ongoing process can be denoted by only one form, namely the *imparfait form*, *entrait*. This form will denote an ongoing process, because it involves the following statement: "X produced an activity with the purpose that it be sufficient for the fact that X exists inside L." This statement is composed of two assertions, but only one concerns the real world, namely, X produced an activity with a certain intention—the other assertion concerns an imagined world, namely, X exists inside L. All three grammatical forms are dicentic signs when they are part of an utterance: they all point to situations in reality by virtue of their potential assertive abilities combined with their temporal value (past) and their modal value (real world).

The natural question arises: Why does the French language has at its disposal three past tense forms among which two forms assert the same and refer to the same event—they represent the same dicent. Why should a language have two forms that assert the same and therefore refer to the same? My answer is that they form a group of three members, each of which denotes one of the three possible forms of argument, that is, abduction, induction and deduction. They do so by means of other verbal categories with the result that any finite verb in the past related tenses of French forms an argumentative sign (cf. table 3). I emphasize that I am talking about *logical argumentation*, which is hearer-oriented in the sense that the speaker informs the hearer by arguing for his or her information. I am not talking about purely speaker-oriented *reasoning* in which the speaker uses abduction, induction and deduction as a heuristic tool to gain information about reality. Nor am I talking about *verification* in which deduction and induction are used as reality-oriented verification methods. The first-person oriented reasoning, the third-person oriented verification and the second-person oriented logical argumentation should be kept apart, because they are quite distinct. If this is so, they must also exhibit different relationships to Firstness, Secondness and Thirdness: Abduction will belong to Firstness when reasoning is concerned, but to Secondness when logical argumentation is concerned—in the case of verification abduction has no place.

7.2. Introducing the Three Logical Argumentation Forms
A complete logical argument (syllogism) consists of two premises and one conclusion drawn from the premises. The three obligatory components are the Rule, Case and Result. This allows us to see what is going on when we shift from one form of argument to another. Since the two premises always appear in pairs, we get three logical types of argument, each involving its own conclusion. This means that Rule, Case and Result can all act as the conclusion, when the remaining two components are the premises. Let us examine the three forms by using an illustrative example (borrowed from Sebeok, 1994, see under *abduction*).

- Form: Deduction—a conclusion going from Rule via Case to Result
- Time: Going forward in time and making a prediction
- Perspective: Prospective

RULE: All the beans from this bag are white
CASE: These beans are from this bag

RESULT: These beans are white

The Result "These beans are white" constitutes the point of departure for the next logical argumentation form:

Form: Abduction—an explanation going from Result via Rule to Case
Time: Going back in time in order to find the answer to the reason why
Perspective: Retrospective

RESULT: These beans are white
RULE: All the beans from this bag are white

CASE: These beans are from this bag

The Case "These beans are from this bag" makes up the starting point for the last logical argumentation form:

- Form: Induction—a generalization going from Case via Result to Rule, not once, but over and over again
- Time: Going forward and then backward in time, from Case to Result and then from Result to Case, to find a law
- Perspective: "Interspective"

CASE: These beans are from this bag
RESULT: These beans are white

RULE: All the beans from this bag are white

The Rule "All the beans from this bag are white" forms the input to the first logical argumentation form, cf. above. In other words, we get a non-ending circle. It is crucial to note that behind the general statement "All the beans from this bag are white" lies a series of single conditional statements: If this bean is from this bag, then it is white; if this bean, too, is from this bag, then is white, and so on and so forth. In other words, although the Rule "All the beans from this bag are white" seems to be one single statement, it is a conditional statement where the antecedent is composed of what was called the Case, that is, (if) the beans are from this bag, and where the consequent is composed of what was called the Result, that is, (then) the beans are white. Let us call the Case *p* and the Result *q*, and say that the Rule is "if p, then q." This clearly shows that the Rule is made up of the Case and the Result—the test "if p, then q" should, however, be multiplied with the amount of beans in the back. In short, the two autonomous descriptions, *p* and *q*, are contained within the entire description, "if p, then q," but appears linguistically as a compact mass which seems to be impossible to decompose, that is, "All the beans from this bag are white." All this is extremely important from our linguistic point of view, since action verbs consist of an activity description and a state description (cf. above).

I shall argue that the *imparfait* together with other verbal categories involves a deductive inference, the *passé composé* an abductive inference and the *passé simple* an inductive inference. In the case of logical argumentation, induction is thirdness: It is a generalization; one goes from a source to a goal and from the goal to the source; and one has an interspective view—with respect to form, time and perspective induction presupposes the two other forms of argument (cf. above). If this is true, we should be able to conclude that the *passé simple* form (in other languages called the aorist) cannot exist alone—the form represents Thirdness that presupposes Secondness and Firstness. In other words, the *passé simple* presupposes the existence of the *imparfait* and the *passé composé*, and not the other way around. This is confirmed by the history of individual languages: It always comes as the last form in the evolution of a language and it always disappears as the first (for further details, see Durst-Andersen, 2006, 2008b). As a matter of fact, what I am arguing at this place is happening just now in front of us: The French *passé simple* has disappeared from oral discourse and is only part of the verbal system of written discourse.

7.3. The Three Past Tense Forms of French as Argumentative Forms
In the following I will try to demonstrate that the *passé simple, passé composé* and *imparfait* should be treated as the linguistic counterparts to the three logical forms of argumentation, that is, induction, abduction and deduction. As pointed out above, all three forms of syllogistic argumentation involve Law, Case and Result but differ with

respect to the way in which the exact argument is built up. If we transfer this to our linguistic forms, they should be composed of two premises that together lead to a valid conclusion. This possibility is only found in so-called action verbs involving an underlying statement of the form p-and-q. Activity verbs, for example, *pleurir* (cry), involve only an activity description *p* and state verbs, for example, *avoir* (have), only a state description *q*. Only action verbs, for example, *entrer* (enter), contain an activity description (p) and a state description (q) that taken together can form a conclusion. Let us test this hypothesis. We shall start with the *passé simple*.

7.3.1. Passé Simple—Induction: Presentation of the Event in Its Totality (p and q)
The French *passé simple* is traditionally said to present the event referred to in its totality (cf. Sten, 1952; Vikner, 1986; Waugh, 1990; etc.):

1. *Jean entra dans la chambre.* (Jean entered the room.)

Entrer is an action verb naming an activity (p) as well as a state (q). This is a precondition for the *passé simple* form to participate in this kind of argumentative structure. State and activity verbs create only a mono-propositional structure and the possibility for entering into an argumentative structure is therefore squeezed out in advance (this is the reason why the *passé simple* of state and activity verbs simply have another meaning—they act as inchoatives or completives). From the point of view of reference, the *passé simple* will always present an action as an event—this is the aspectual meaning of the form that relates to the RULE in the argumentative structure. Its temporal meaning is past tense—both the activity and the state referred to by the form belong to the past world. This type of content is reflected in the two premises, that is, the CASE (the activity description) and the RESULT (the state description). We have now accounted for all pieces of content except for one, namely the presentational properties of the form.

   How is it possible to account for the fact that the *passé simple* presents the event in its totality, as it is suggested by all existing theories of the French *passé simple*. It is exactly here that the logical-narrative properties of the form come into the picture. It is not only stated that the activity description p and the state description q constitute the premises, but also that the two descriptions are presented as one bloc, that is, as a message in which the very description of (p causes q) is presented holistically, that is, as one information unit in its totality (if p, then q). In other words, the form transforms (p causes q) in a past world into a condensed message. The essence of induction has to do with that fact that one makes a generalization from a Case (e.g., "These beans are from this bag") and a Result (e.g., "These beans are white") by putting these two isolated descriptions into one single description without losing its content at all (e.g., "All the beans from this bag are white," which means: If the beans are from this bag, then they are white). The two autonomous descriptions are contained within the entire description, but as a compact mass. This is exactly the same compactness we find in the French *passé simple* and therefore also the argumentative direction of induction, that is, (if *Jean produced an activity of entering, then Jean existed in the room*) is presented *en bloc* as *Jean entered the room*:

Grammatical argumentation: Induction—Setting up a rule/generalization
    CASE: Jean produced an activity (p)
    RESULT: Jean existed in the room (q)

    RULE: if p, then q

It is interesting that this description does not add anything new to what has hitherto been said by other theories (see, for instance, Nolke & Olsen, 2003; Waugh, 1975, 1990; Vikner,1985; Sten, 1952): They all claim that *passé simple* views an action in its totality. What is new is that the theory explains why it is so and how it is achieved.

The hearer knows that the *passé simple* involves an inductive form of argumentation and he or she is therefore capable of reconstructing the situations behind the message, namely that Jean produced an activity and that this activity caused a state where Jean exists inside a certain room—the effect of the activity. This reconstruction is made by the hearer and is not part of the meaning of the form. In other words, the *passé simple* is the linguistic analogue to motion picture in which the causal activity and the resultant state melt together, but the form itself only conveys information. If you wish to say the same by using a motion picture presentation you will have to use the *imparfait* (the use is normally called *imparfait pittoresque* [the imperfect as picture description]—this is often done, for instance, when reporting directly from soccer matches; cf. Berthonneau & Kleiber, 1999; Haff, 2005).

7.3.2. Passé Composé—Abduction: A Present State (q) is a Consequence of a Past Activity (p)
The French *passé composé* is used to give an explanation in information terms:

2. *Jean est entré dans la chambre.* (Jean has entered the room.)

The *passé composé* also belongs to the perfective aspect, since it refers to an event—in other words, both the *passé simple* form and the *passé compose* form denote the same—from the point of view of the dicisign they are identical. Both forms imply that Jean produced an activity that was sufficient for Jean's presence in the room. They differ, however, with respect to the way in which they present the past event. Unlike the *passé simple*, the *passé composé* has the state description q as foreground and the activity description p as background. *Est* in *est entré* is simply an index of this foregrounding—it forms the first premise—while *entré* in *est entré* is an index of the past activity to which the speaker returns in his conclusion. In other words, the *passé composé* asserts that "Jean exists inside the room" is true in a present world. Moreover, the truth of the state description is explained by the truth of the activity description "Jean produced an activity," which holds good in a past world. In other words, the *passé composé* does not only give a news-flash, but at the same time it gives an explanation in information terms:

Grammatical argumentation: Abduction – Explanation

> RESULT: Jean exists in the room (q)
> RULE: If p, then q
> ----------------
> CASE: Jean produced an activity (p)

Due to the fact that *entrer* does not specify the activity itself, the rule may apply to any possible world in which Jean either flew, ran, swam or walked into the room. It should be noted that the *passé composé* is not only used in its news-flash function (which requires an action verb): It is also used in connection with state and activity verbs to give a characterization of a person with the same argumentative structure: *J'ai habité à New York pendant deux ans* (I have lived two years in New York [but I am not living there anymore]) has exactly the same meaning: I have the quality of having lived in New York for two years because "I lived in New York two years." In other words, the *passé composé* gives a characterization of a person by offering a physical explanation of a psychological quality which will hold good for the rest of that person's life. Once again the abductive inference is quite evident. The Rule is here that all physical and psychological experiences from the past inhere in the person as present time qualities. In other words, the *passé composé* is a hybrid form. As an aspectual form it belongs to the perfective aspect because it presents an action as an event (cf. the RULE). As a temporal form it is a present form (cf. the RESULT) as well as a past form (cf. the CASE) because it describes the present world as well as the past world. And as a logical-narrative form it has the argumentative direction which is found in abduction: A present state is explained by a past situation via a rule.

7.3.3. Imparfait—a past activity (p) having a consequence (q)
The French *imparfait* is used to give a situation description as well as a characterization in connection with state and activity verbs:

3. *Jean faisait le taxi* (characterization)*, pendant qu'il terminait* (situation description) *ses études.* (Jean drove [characterization] a taxi [i.e., he was a taxi driver], when he was finishing [situation description] his studies.)

Since a state verb names *q* and an activity verb names *p*, there is no possibility to enter into an argumentative structure, which requires two premises. Only in connection with action verbs, which name p as well as q, that possibility arises. Here we find the major function of the *imparfait*, where it is used to present an action as an ongoing process, that is an activity having a state as its purpose:

4.   (*Comme*) *Jean entrait dans la chambre* (*, il vit Marie*). (Jean was entering the
     room [when he saw Marie]).

The French *imparfait* has the activity as its foreground and the state as its
background—it is the converse of the *passé composé*. As an aspectual form it belongs
to the imperfective aspect, which means that it is directly opposed to both the *passé
composé* and the *passé simple*. The form is multi-ambiguous (see, for instance, Engel
& Labeau, 2005; and Labeau, 2002 where all its contextual meanings proposed by
various aspectologists are given), but for our purpose the following specific meanings
will suffice:

5.   *Il lisait un livre* [Ongoing process]. (He was reading a book.)

6.   *Une heure plus tard, il prenait le train pour Paris* [Planned action]. (An hour later
     he was going to take the train for Paris.)

7.   *Une heure après, il prenait le train pour Paris* [Imparfait pittoresque]. (An hour
     before he had taken the train for Paris.)

8.   *Il tombait de fatigue* [Imparfait de conatu]. (He was about to fall from fatigue).

9.   *Si tu avançais, je frappais* [Imagined world]. (If you step forward, I will hit you.)

This suggests that the *imparfait* is the unmarked member: it may refer to an
ongoing process (cf. 5), a planned action (cf. 6), an event (cf. 7), a non-event (cf. 8)
and a future event (cf. 9). What remains common to all these uses is the descriptive
function. The *imparfait* fulfills this function by asserting that the activity description
(p) is true, and by leaving the state description (q) as what I call a standard
implicature, that is, it is up to the hearer to decide whether it is true or false in the
situation referred to. Sometimes the state description is false (cf. 5, 6 and 8),
sometimes it is true in a real world (cf. 7) or in an imagined world (cf. 9). As a tense
form it places the activity as well as the state in the past world. As a logical-narrative
form it gives the argumentative direction found in deduction:

Grammatical argumentation: Deduction – Conclusion

     RULE: if p, then q
     CASE: Jean produced an activity (p)

     RESULT: Jean was in the room (q)

The speaker takes his starting point in a Rule that says that if an activity
description is true, then the state description is true, if we are dealing with all normal
worlds—in an abnormal world people may fall and break their leg when entering a

room. By presenting the hearer with a past activity without specifying the state, the hearer is forced to decide whether the state will obtain at a certain later moment. If we are dealing with a normal world, the hearer will conclude that the state does not obtain at the moment of reference, but will obtain at a certain later moment. Note that (7) is always used as a conclusion and that (9) can be said to repeat the argumentative direction of the deductive inference, but to remain in a logical world, that is, an imagined world. In that way this logical approach is also capable of explaining these two interesting uses of the *imparfait*.

7.4. Concluding Remarks

The paper has shown that the three forms that make up the fundamental aspect-tense-mood system of French seem to conform to the three argumentative forms proposed by Peirce. Interestingly, this theory adds nothing new to the description and use of the three forms, but it explains how each of them functions as they do. The idea that deduction, abduction and induction form an integrated and substantial part of grammar is interesting from the point of view of the history of mankind. From that point of view, Peirce only found what was already tacitly formulated and agreed upon by members of a speech community. My hypothesis is that deduction, abduction and induction basically reflect three ways of producing motion: a) deduction—X goes from Location 1 to Location 2; b) abduction—X goes from Location 2 back to Location 1; and c) induction—X goes from Location 1 to Location 2 and back again (repeatedly). The three ways of producing motion are then applied to (1) visual perspective (e.g., the extrovertive, the introvertive and correlative viewpoints, cf. Durst-Andersen, 1996), (2) reasoning, (3) verification, and (4) logical argumentation (cf. Durst-Andersen, 2008b).

8. Concluding Summary

The main conclusions of this paper are straight forward. First, linguistics was never driven as a true semiotic science. Its growth as a semiotic science was not facilitated, but hindered by Saussure's claim that the linguistic sign is arbitrary. People concluded that because the relation between content and expression of the sign is arbitrary, its way of expressing it is not significant at all. One forgot that the only way in which languages can show their differences is by having different forms of expression. Secondly, in relying on Morris' three dimensions of the linguistic sign, namely syntax, semantics and pragmatics, linguists believed that they were dealing with semiotics, but, as a matter of fact, they remained at the periphery of linguistic semiotics but never got into its centre. Thirdly, linguistics, communication science and pragmatics proper were never driven by focusing equally on the three participants of a communication situation. They were exclusively centred on either (a) third person properties (e.g., Glossematics), (b) first person properties (e.g., transformational grammar and functional grammar) or (c) second person properties (e.g., relevance theory). By ignoring one or two of the obligatory three participants, that is, reality, speaker and hearer, it was completely impossible to reach the true level of the dialogue (cf. Bakhtin, 1929/1994).

I took my starting point in Peirce's thrichotomic definition of a sign and defined the lexeme (a simple sign) as an expression unit that mediates an image-idea pair. This enabled me to account for the relationship between perception and cognition on the one hand and the various lexicalization patterns found in different languages, on the other. It was demonstrated that any lexeme is static and completely helpless in a communication situation, because it should be omnipotent, that is, capable of being used in any situation. This omnipotence is ensured by its arbitrariness, which in no way should be confused with lack of motivation. Due to its static character any lexeme needs a grammar that can give it a semiotic direction. This dynamism is achieved at the level of utterance (a complex sign), which functions as a prime index, that is, any utterance points at something. Since what is normally called a state of affairs has three modalities of existence in any concrete communication situation, that is, the situation as such, the speaker's experience of it and the hearer's experience of it, members of a speech community must agree upon the semiotic direction: Will they talk about situations being common to the speaker as well as the hearer, or about the speaker's experiences, or will they compare their own experiences with those of the hearers and, as a result of that comparison, present various pieces of information to the hearer? The choice cannot be avoided, because a grammar would not function, if it used all three semiotic directions at the same time. A language must choose between three types of indexes (or indices), that is, model, symptom or signal. But, although any grammar has to make a choice whereby the two other alternatives are ignored, the hearer makes up for this during his decoding process. He or she knows that any utterance will be a model of a situation, a symptom of the speaker's experiences and a signal to the hearer and will therefore extract the lacking pieces of information in order to make the incomplete message complete. This is the reason why speakers of different types of languages get almost the same out of an utterance—their grammar speaks with one voice, but the hearer compensates for this by filling out the two other voices. In that way any utterance ends up being polyphonic.

The paper discussed another important aspect which does not seem to be limited to language: how inherent properties of a sign become external functions. According to Peirce, any symbol is built upon an index, which, in turn, is built upon an icon. But the symbol itself has become an abstract unit that has to be wrapped out in order to be properly understood. This is ensured, on the one hand, by the integrative function of grammemes—they act as indexes, and, on the other, by their differentiating function—they point to the presence or absence of iconicity. In other words, the two functions of grammemes seem to repeat the inherent properties of any symbol, but in the opposite order of appearance.

Traditionally, grammemes have been grouped into pairs, which are analyzed in binary terms. This approach has been successful in many cases, but it became extended to the analysis of all verbal forms—also to cases in which it is obvious that we are dealing with three members of a category. Such forms cannot be analyzed in binary terms but should be studied as hierarchies. This concerns, for instance, the *passé simple, the passé compose* and the *imparfait* in written French discourse. It was demonstrated that they form a group of three members, because together with other verbal categories the three forms are used to present the hearer with the argumentative forms of induction, abduction and deduction, respectively. These logical forms are normally considered to be text functions—which they are—but it is important that they can be seen as exteriorizations of the inherent lexical-grammatical properties of the French verb. If this is true, we are facing an interesting phenomenon: Within the verb itself we witness a polyphonic internal dialogue.

Acknowledgements

I want to express my deepest gratitude to Winfried Nöth for his major effort to make this paper better. I also want to express my warm thanks to Marcel Danesi for all his inspiring comments and criticism.

References

Andersen, H. (2001). Markedness and the theory of linguistic change. In H. Andersen (Ed.), *Current issues in linguistic theory: Vol. 219. Actualization* (pp. 19-57). Amsterdam: John Benjamins.

Bakhtin, M. M. (1994). *Problemy tvorčestva Dostoevskogo.* Kiev: Next. (Originally published in 1929 by Priboj in Leningrad)

Baron, I. (2005). Aspects lexicaux de la traduction des noms composés. *Voprosy filologii,* (3), 62-68.

Beaugrande, R. de. (1985). Text linguistics in discourse studies. In T. A. Van Dijk (Ed.), *Handbook of discourse analysis, Vol. 1* (pp. 41-70). London: Academic Press.

Benveniste, É. (1966). *Problèmes de linguistique générale.* Paris: Gallimard.

Berthonneau, A.-M., & Kleiber, G. (1999). Pour une réanalyse de l'imparfait de rupture dans le cadrede l'hypothèse anaphorique méronomique. *Cahiers de praxématique* [Montpellier], *32,* 119-66.

Bhat, D. N. S. (1999). *The prominence of tense, aspect and mood.* Amsterdam: John Benjamins.

Bloomfield, L. (1935). *Language.* London: Allen & Unwin.

Bühler, K. (1934). *Sprachtheorie. Die Darstellungsfunktion der Sprache.* Stuttgart: Gustav Fischer Verlag.

Chomsky, N. (1957). *Syntactic structures.* The Hague: Mouton.

Chomsky, N. (1981). *Lectures on government and binding.* Dordrecht: Foris.

Chomsky, N. (1986). *Knowledge of language: Its nature, origin, and use.* New York: Praeger.

Contini-Morava, E., Kirsner, R. & Rodriguez-Bachiller, B. (2004). *Studies in Functional and Structural linguistics: Vol. 51. Cognitive and communicative approaches to linguistic analysis.* Amsterdam: John Benjamins.

Coseriu, E. (1971). Synchronie, Diachronie und Typologie. In *Sprache, Strukturen und Funktionen* (pp. 91-106). Tübingen: Gunter Narr Verlag. (Originally published in 1965 as "Sincronia, diachronia y tipologia" in Actas del XI Congeso Internacional de lingüística y filología Románicas, Madrid 1965, 1 Vol., 269-281.)

CP: See Peirce, C. S. (1931-35).

Devitt, M. & Sterelny, K. (1987). *Language and reality: An introduction to the philosophy of language.* Oxford: Basil Blackwell.

Dodge, E., & Lakoff, G. (2005). Image schemas: From linguistic analysis to neural grounding. In B. Hampe (Ed.) *From perception to meaning: Image schemas in cognitive linguistics* (pp. 57-91). Berlin: Mouton de Gruyter.

Dik, S. (1989). *The theory of functional grammar. Part one: The structure of the clause.* Dordrect: Foris Publications.

Durst-Andersen, P. (1992). *Mental grammar: Russian aspect and related issues.* Colombus, Ohio: Slavica Publishers.

Durst-Andersen, P. (2005a). Mood and modality in Russian, Danish and Bulgarian: Determinant categories and their expanding role. In A. Klinge & H. Høeg Müller (Eds.), *Modality: Studies in form and function* (pp. 215-246). London: Equinox.

Durst-Andersen, P. (2005b). *Obščie i specifičeskie svojstva grammatičeskix sistem. K postroeniju novoj teorii jazyka* [The general and the specific features of grammatical systems. Towards a new theory of language]. Moscow: RGGU.

Durst-Anderssen, P. (2006). From propositional syntax in Old Russian to situational syntax in Modern Russian. In O. N. Thomsen (Ed.), *Competing models of linguistic change: Evolution and Beyond* (pp. 211-234). Amsterdam: John Benjamins.

Durst-Andersen, P. (2008a). Linguistics as semiotics: Saussure and Bühler revisited. *Signs, 2*: 1-29.

Durst-Andersen, P. (2008b). The two aspectual systems in French. In M. Birkelund, M-B. M. Hansen & C. Norén (Eds.), *L'énonciation dans tous ses états. Mélanges offerts à Henning Nølke à l'occasion de ses soixante ans* (pp. 373-494). Bern: Peter Lang.

Eco, U. (1975). *A theory of semiotics.* Bloomington, IN: Indiana University Press.

Engel, D. M., & Labeau, E. (2005). Il était une fois un match de foot: L'événement sportif comme objet de narration. *Revue Romane, 40*, 199-218.

Gerlach, C. (2008). *Visual object recognition and category-specificity.* Post-doctoral dissertation. Copenhagen: Dept, of Psychology, Copenhagen University.

Haff, M. H. (2005). L'imparfait narratif – l'enfant terrible de l'univers aspectuo-temporel francais. *Revue romane, 40*, 137-52.

Halliday, M. A. K. (1975). *Learning how to mean.* London: Arnold.

Hengeveld, K. (2004). The architecture of a Functional Discourse Grammar. In J. Lachlan Mackenzie & M. Á. Gómec-González (Eds.), *Functional Grammer Series: Vol. 24. A new architecture for Functional Grammar* (pp. 1-21). Berlin: Mouton de Gruyter.

Herslund, M., & Baron, I. (2003). Language as world view. Endocentric and exocentric representations of reality. In I. Baron (Ed.), *Copenhagen Studies in Languages: Vol. 29. Language and culture* (pp. 29-42). Copenhagen: Samfundslitteratur.

Hjelmslev, L. (1943). *Omkring sprogteoriens grundlæggelse.* Copenhagen: Munksgaard.

Hockett, C. F. (1958). *A course in modern linguistics.* New York: MacMillan Publishing Co., Inc.

Hockett, C. F. (1963). The problem of universals in language. In J. H. Greenberg (Ed.), *Universals in language* (pp. 1-22). Cambridge, MA: The MIT Press.

Jakobson, R. (1960). Linguistics and poetics. In T. A. Sebeok (Ed.), *Style in language* (pp. 350-377). Cambridge, MA: The MIT Press.

Jakobson, R. (1962). Zeichen und System der Sprache. In *Selected Writings* II (pp. 272-279). The Hague: Mouton.

Jakobson, R. (1963). Efforts toward a means-end model of language in interwar Continental linguistics. In *Selected Writings* II (pp. 522-526). The Hague: Mouton.

Jakobson, R. (1965). Quest for the essence of language. In *Selected Writings* II (pp. 345-359). The Hague: Mouton.

Jakobson, R. (1968). Language in relation to other communication systems. In *Selected Writings* II (pp. 697-708). The Hague: Mouton.

Jakobson, R. (1971). Retrospect. In *Selected Writings* II (pp. 709-722). The Hague: Mouton.

Jensen, M. R. (1999). *Copenhagen Working Papers in LSP: Strukturen i russiske gestus. En semiotisk, semantisk og pragmatisk analyse.* Copenhagen: Handelshøjskolen i København.

Kendon, A. (1995). Gestures as illocutionary and discourse structure markers in Southern Italian conversation. *Journal of Pragmatics, 23*, 247-279.

Krampen, M. (1986). Ontogenesis of iconicity and the accessibility of the mental image by children's drawings. In J. D. Evans & A. Helbo (Eds.), *Semiotics and international scholarship: Towards a language of theory* (pp. 85-99). Dordrecht: Martinus Nijhoff Publishers.

Krampen, M. (1990). *Children's drawings. An iconic coding of the environment.* New York: Plenum Press.

Labeau, E. (2002). L'Unité de l'imparfait: vues theoretiques et perspectives pour les apprenants du francais langue étrangère. *Traveaux de linguistique, 45*, 157-184.

Lakoff, G. (1987). *Women, fire, and dangerous things: What categoires reveal about the mind.* Chicago: University of Chicago Press.

Langacker, R. W. (1999). *Grammar and conceptualization.* Berlin: Mouton de Gruyter.

Langacker, R. W. (1991). *Foundations of cognitive grammar. Vol. 2: Descriptive application.* Stanford, CA: Stanford University Press.

Lucid, D. P. (Ed.) (1977). *Soviet semiotics. An anthology.* Baltimore: The Johns Hopkins University Press.

Lyons, J. (1977). *Semantics. Vol. 1.* Cambridge: Cambridge University Press.

Martinet, A. (1949). La double articulation linguistique. *Travaux du Cercle Linguistique de Copenhague, 5*, 30-37.

Morris, C. W. (1938). *International encyclopedia of unified science: Vol. 1* (2). *Foundations of the theory of signs* Chicago: The University of Chicago Press.

McNeill, D. (1992). *Hand and mind. What gestures reveal about thought.* Chicago: University of Chicago Press.

Nolke, H., & Olsen, M. (2003). Le passé simple subjectivisé. *Langue Francaise, 138*, 5-85.

Nöth, W. (2002). Charles Sanders Peirce, pathfinder in linguistics. *Interdisciplinary Journal for Germanic Linguistics and Semiotic Analysis, 7*, 1-14.

Parker, Kelly A. (1997). *The continuity of Peirce's thought.* Nashville: Vanderbilt University Press.

Peirce, C. S. (1931–1935, & 1958). *The collected papers of Charles Sanders Peirce.* Vols. I–VI [C. Hartshorne & P. Weiss, Eds., 1931–1935], Vols. VII–VIII [A. W. Burks, Ed., 1958]. Cambridge, MA: Harvard University Press. (Citations use the common form: CP vol.paragraph).

Peirce, C. S. (1953). Charles S. Peirce's letters to Lady Welby (I. Lieb, Ed.). New Haven: Yale University Press.

Peirce, C. S. (1958). *Reviews, correspondence and bibliography. Collected Papers of Charles Sanders Peirce* (Vol. VIII, A. W. Burks, Ed.). Cambridge, MA: Harvard University Press.

Peirce, C. S. (1998). *The Essential Peirce: Selected Philosophical Writings. Volume 2 (1893-1913)* (Peirce Edition Project, Eds.). Bloomington, IN: Indiana University Press. (Citations use the form: EP vol., page).

Piaget, J., & Inhelder, B. (1966). *L'image mentale chez l'enfant.* Paris: Presses Universitaires de France.

Ransdell, J. (1977). Some leading ideas in Peirce's semiotic. *Semiotica, 19*, 157-178.

Ritchey, G. M. (1980). Picture superiority in free recall: The effects of organization and elaboration. *Journal of Experimental Child Psychology, 29*, 460-74.

Saussure, F. de. (1916). *Cours de linguistique générale*. Paris: Payot.

Searle, J. R. (1969). *Speech acts: An essay in the philosophy of language*. Cambridge: Cambridge University Press.

Sebeok, T. A. (Ed.) (1994). *Encyclopedic dictionary of semiotics*. Berlin: Mouton de Gruyter.

Sebeok, T. A., & Danesi, M. (2000). *The forms of meaning: Modeling systems theory and semiotic analysis*. Berlin: de Gruyter.

Serebrennikov, B. A. (Ed.) (1973). *Metody lingvističeskix issledovanij*. Moscow: Akademija nauk.

Shapiro, M. (1983). *The sense of grammar: Language as semiotics*. Bloomington, IN: Indiana University Press.

Shapiro, M. (2003). Aspects of a neo-Peircean linguistics: Language history as linguistic theory. In M. Shapiro (Ed.), *The Peirce seminar papers, Vol. V* (pp. 108-125). Oxford: Berghahn Books.

Sheriff, J. K. (1994). *Charles Peirce's guess at the riddle. Grounds for human significance.* Bloomington, IN: Indiana University Press.

Short, T. L. (2004). The development of Peirce's theory of signs. In C. Misak (Ed.), *The Cambridge companion to Peirce* (pp. 214-40). Cambridge: Cambridge University Press.

Sperber, D., &Wilson, D. (1986). *Relevance: Communication and cognition.* Oxford: Basil Blackwell.

Sperber, D., &Wilson, D. (2004). Relevance theory. In L.R. Horn & G. Ward (Eds.), *The handbook of pragmatics* (pp. 607-632). Oxford: Basil Blackwell.

Sten, H. (1952). *Les temps du verbe fini (indicatif) en francais moderne*. Copenhagen: Munksgaard.

Talmy, L. (2001). *Toward a cognitive semantics. Vol. II: Typology and process in concept structuring*. Cambridge, MA: The MIT Press.

Vikner, C. (1985). Aspect in French: The modification of aktionsart. In F. Sørensen (Ed.), *Aspects of aspect* (=CEBAL SERIES, 9, pp. 58-101). Copenhagen: Nyt Nordisk Forlag Arnold Busck.

Voloshinov, V.N. (1986). Marxism and the philosophy of language (L. Matejka & I. R. Titunik, Trans.). Cambridge, MA: Harvard University Press.(Originally published in 1929 by Priboj in Leningrad)

Waugh, L. R. (1975). A semantic analysis of the French tense system. *Orbis, 24*, 436-85.

Waugh, L. R. (1990). Discourse functions of tense-aspect in French: Dynamic synchrony. In N. Thelin (Ed.), *Verbal aspect in discourse* (pp. 159-190). Amsterdam: John Benjamins.

Wierzbicka, A. (1980). *Lingua mentalis: The semantics of natural language*. Sydney: Academic.

Craft, D. (2009). *Plasticity Index No. 2*. Microphoto: Carbolic Acid Melt Crystals; 34 x 21 cm.

Craft, D. (2009). *Vortex No. 2*. Microphoto: Carbolic Acid Melt Crystals; 34 x 21 cm.

Cybernetics and Human Knowing. Vol. 16, nos. 3-4, pp. 81-88

Biosemiotics:
To Know, What Life Knows

Kalevi Kull[1]

The field of semiotics is described as a general study of knowing. Knowing in a broad sense as a process that assumes (and includes) at least memory (together with heredity), anticipation, communication, meaningful information, and needs, is a distinctive feature of living systems. Sciences are distinguished accordingly into phi-sciences (that use physicalist methodology) and sigma-sciences (that use semiotic methodology). Jesper Hoffmeyer's book Biosemiotics is viewed as an inquiry into the sigma-scientific approach to living systems.
Keywords: Biosemiotics, knowing, adaptation, Φ-sciences, Σ-sciences, Hoffmeyer

Whether biology has studied what organisms know?

Lamarck, after his countrymen, was probably one after whom biology took its shape as a study of adaptations—both the processes of adaptation and the organic forms as adaptations. Once the correspondences of organisms to the world are not eternal but a result of (ontogenetic or phylogenetic) experience, the adaptations turn out to be a kind of knowledge, a model that is acquired or worked out via certain mechanisms (either developmental or evolutionary). Possessing an adaptation or habit would mean that one has some experience through which the adaptation has formed.

Thus via a study of organic functions that characterize adaptations, biology has described the information the organisms have, including the memory and the purpose. Adaptations which are always relations, bonds of life, qualitative phenomena, can be seen as iconic relations (or in more complex situations also as indexical or symbolic relations), that is, sign relations. Since the sign relations are modelling relations, so are adaptations certain kinds of models.[2] Their description and explanation has turned the attention of biology towards the concepts of history and meaning. And this has always made biology the "humanities" of the natural sciences.

Stating this, one should notice that much of the work in biology has digressed from this path. Once the adaptation is defined quantitatively via fitness—via the number of copies one makes—its fundamental feature of qualitative *fit* is lost from the description. In other words, the meaning was lost. This is like description of sign without paying attention to its reference.

Back on the track with a semiotic approach, it simply means that we are going to take the biological study of qualitative relations seriously. In this, Jakob von Uexküll's approach has been most steadfast. However, at least as a tacit intention, biology has

---

1. Department of Semiotics, University of Tartu, Estonia
2. Although, there have not been many writings in which a morphological adaptation has been directly identified as a *model* produced by the organism (e.g., Frost, 1987).

obviously always attempted to describe the billion-year experience that living systems hide in themselves.

1. Semiotics and life's knowing

What makes the living and the non-living very different from each other is their different relation to what is not, to what is absent. That which is alive has expectancies. Once dead, one does not expect.

Expectancy, or anticipation, means that something is not only itself but also stands for something else – for that which is expected. This is exactly the general feature of the sign, according to Peirce's widely repeated definition.

The sign vehicle, or representamen (or sign, *sensu stricto*), stands for an object. This is the relation that is created by semiosis. The object, thus, has an interesting duality—it is both there and is not there—because it is both connected and anticipated.[3] The relation of *standing for* is possible owing to the absence of what is referred to (the object) and, concurrently, there cannot be semiosis without the existence of a reference (an object).

Analogously *not being something* may characterize also the sign vehicle, and interpretant—this is the very feature of being a sign. Semiosis is what makes anything plural.[4] Semiosis creates objects, and makes each object plural. Each is sign, which means each is simultaneously something else, each is many. To mean is to be plural.

The simultaneous *what-is* and *what-is-not* is related to another fundamental feature of the sign—simultaneous arbitrariness and non-arbitrariness. In its *standing for*, the sign's relation to object is always arbitrary (because one has not determined the other). As a relation to the object in the triadic relation, the object exists in an entirely non-arbitrary way.

How such a duality (or more exactly, plurality) occurs, can be described by the models of semiosis, of which the Peircean triadic model and the Uexküllean functional cycle model are currently the two most valuable.[5] Both descriptions include the epistemological aspect of the interpretation process and thus disclose the ontological plurality of meaning.

Semiosic systems are simultaneously modelling systems, as was emphasized already in the 1960s by the Tartu-Moscow school (Lotman, 1967; Levchenko & Salupere, 1999). Anderson and Merrell (1991, p. 4) admit that "signs ... are themselves models, and semiosis constitutes modeling, par excellence." The modelling systems include both the organismic and cultural systems. Thus, semiotics can be "modeling of modeling" (Anderson & Merrell, p. 4).

This essential epistemological aspect of semiosis—that semiosis is a modelling process—has been emphasized not only by Tartu school. Sebeok's (2001, p. 156) view strongly resonates: "*semiosis* [is the] capacity of a species to produce and

---

3. On the concept of anticipation, see also Kull (1998).
4. See also Kauffman (2005), Kull (2007b), Merrell (2007).
5. See also Krampen (1997).

comprehend the specific types of models it requires for processing and codifying perceptual input in its own way." Via Sebeok's thesis, which identifies life process as semiosis, also life process turns out to be modelling.

Robert Rosen, concluding his major work, *Life Itself*, writes in the final paragraph of the book: "We began this discussion with the question, 'What is life?' We ended with the answer: Life is the manifestation of a certain kind of (relational) model. A particular material system is *living* if it realizes this model" (Rosen, 1991, p. 254). If we use the word *model* in the broadest sense, then we could also say: Life is the process of modelling *sensu lato*, the living system is itself a model.

This aspect of life can be specially focused, which means that organisms can be studied from the point of view of a theory of knowing—that is, of semiotics, including biosemiotics. This is because semiosis—that is life process—does not only transfer messages, it also produces messages, that is, a knowledge.[6]

The methodology of empirical biosemiotic work, however, is not that of the physical sciences, nor of the humanities or the social sciences in which one can carry out interviews.[7] The main question of biosemiotic methodology is: *how to know what organisms know*. Similarly, Robert Rosen has stated that "what is important in biology is not how we see the systems which are interacting, but how they see each other" (Rosen, Pattee, & Sormorjai, 1979, p. 87). Empirical studies in biosemiotics should be studies of knowing what is available to other species, and how they obtain it, and how they use it.

In this way, using a very general notion of knowing (the one that precisely covers the sphere of signs, or meaningful information),[8] it would be easier to see the methodological particulars.[9] Keeping in mind that if what we are studying are not molecules, but the *knowing* that living systems possess, it is obvious that, for instance, statistical methods and measurements can only be very much secondary tools in this research. On the other hand, field experiments would be an important tool in order to study what organisms can distinguish; still, though, the principle of *ceteris paribus* is generally inapplicable, because knowledge is relational in principle. This is entirely scientific, despite the fact that it appears to be opposed to the methodological standards of the physical sciences. The way to acquire truth in semiotics is rather the way we utilise dialogue, or implement the process of translation.

---

6. The concept of knowledge has been treated in a similar way by Canguilhem (2008).
7. Cf. Manning (1987).
8. Similarly, Tommi Vehkavaara (1998) has written about the extended concept of knowledge that would characterize living beings.
9. It is relevant to add here that, according to Short (2007, p. 289), "Peirce's early semeiotic was a *theory of mind*: it identified thoughts as signs interpreting signs. The mature semeiotic *retains that identification* but embeds it in a wider context" (my emphasis). "The mature semeiotic entails a naturalistic theory of mind" (Short, p. 289) in two respects — as a natural history of semiosis (including animal behavior), and as attributing an intentionality to all forms of it (e.g., as also Deely, 2007 argues for). Short (p. 289) adds: "The attribution of purpose, although it grounds valuation, is an empirical hypothesis, testable against observation. And the explanation of purposeful behavior, although not mechanistic, is naturalistic."

2. *Σ*-sciences

Semiosic practice—the behavior of organisms, including the everyday behavior of humans—is an acquisition and application of knowledge. In humans, due to the usage of propositional and narrative sign systems, semiosic practice turns (develops) into knowing of knowing, or semiotics *sensu lato* (in the sense of the human as semiotic animal). A methodical acquisition and organization of knowledge leads to science.

Sciences have been developed into diverse kinds. For instance, Peirce makes a distinction between idioscopic (primarily physical) and cenoscopic (primarily semiotic) sciences.[10] Following Locke, sciences can be divided into three: physics, ethics, and semiotics.

> "Science may be divided into three sorts. ... First, The knowledge of things, as they are in their own proper beings, their constitutions, properties, and operations ... This in a little more enlarged sense of the word, I call *Phusike* ... Secondly, *Praktike*, The skill of right applying our own powers and actions, for the attainment of things good and useful. The most considerable under this head, is ethicks ... The third branch may be called *Semeiotike*, or the doctrine of signs"; "the ways and means whereby the knowledge of both the one and the other of these, are attained and communicated." (Locke, 2008, pp. 462–463 [Ch. 21])

The first deals with knowledge as knowledge about things, the second as knowledge about purpose, and the third as knowledge about knowledge. Complementarity between these sciences is obvious; however, the principles of building the theory and the methods of practice (e.g., descriptive, prescriptive, interactive, and appraising) evince vast differences. Without focusing on ethics as science here, in order to place biosemiotics into a relation with physicalist biology, we still need to distinguish between physical theory and semiotic theory as two basic complementary ways of doing science. Accordingly, the models built in science are of two quite different kinds: physical, and semiotic (*sensu stricto*).

More precisely, in philosophy of science, Rein Vihalemm (2007) has pointed to a different role that the historical explanation has in the major types of sciences, and on this bases he has made a distinction between *Φ*-sciences and non-*Φ*-sciences, the latter called also *Σ*-sciences. *Φ*-sciences do not require historical explanation, they model the world using universal laws and depend on quantitative methods; *Σ*-sciences, instead, are dependent on historical explanations, they model the world on qualitative basis and use primarily qualitative methods.

The two are based on (and interrelated via) the semiosic practice, which includes everyday communication and practical processes of classification and measurement (Table 1).[11] As mentioned by Peirce, "measurement ... is a business fundamentally of

---

10. Due to the wording given by Peirce, it is easy to misidentify his distinction. Following Jeremy Bentham, Peirce defines idioscopy as the science "which discovers new phenomena", and cenoscopy as the "science that is founded upon the common experience of all men" (CP 8.199). A bit misleading may be his formulation about cenoscopy as "studies which do not depend upon new special observations" (CP 8.342), which he still directly identifies as science of *semeiotic*, or the cenoscopic science of signs" (CP 8.343). However, I think that the correspondence between Locke's and Peirce's distinctions can be interpreted as correct.

the same nature as classification" (CP 1.275). At the stage of modelling and theory, however, the approaches diverge, Φ-science as a modelling based on quantitative convertibility, and Σ-science as a modelling based on qualitative diversity. An important point here is that this is not the well-known separation between the humanities and the sciences; this is, instead, science as it necessarily includes the complementary approaches. In principle, Φ-sciences can cover the whole world via physical descriptions, and Σ-sciences can cover the whole world as the sum of knowledge. However, it would be weird for Φ-sciences to describe the meaningfulness, whereas Σ-sciences may include at least the science of all forms of life as much as it studies organisms' knowledge and sign processes.

Table 1: The Distinction Between Σ-sciences and Φ-sciences, or Physical and Semiotic Approaches.

SOCIAL (everyday) domain	Semiosic Practice (*inter alia* includes measurement and classification as human activities)	
SCIENTIFIC subdomains	Φ-sciences Physical approach Study of things	Σ-sciences Semiotic approach Study of knowing
General assumptions for MODELS	Universal laws of nature Faultless world Monist ontology	Local codes Erroneous world Pluralist ontology

3. Why Biosemiotics: Jesper Hoffmeyer's Book

Why is it that biology cannot do without a semiotic approach, and why is it that semiotics cannot do without biosemiotics? Jesper Hoffmeyer's (2008)[12] book provides further answers to these questions. The route to the understanding of life itself has been, via many models, meandering and blundering. Hoffmeyer lists reductionism and vitalism, Cartesian dualism and anthropomorphism, etc., and carefully compares the biosemiotic approach with many others on the basis of the analysis of particular biological examples. It does this in a much more detailed fashion than in his earlier *Signs of Meaning in the Universe* (Hoffmeyer, 1996).[13] In a way, Hoffmeyer is like a therapist of biology (likewise, a role of semiotics, in part, can be seen as being a therapist of culture).

A few more words may illustrate the point here. Hoffmeyer (2008, p. 24) writes: "Biosemiotics considers human mental processes not as unique phenomena in the

---

11. Regarding this distinction, see also Kull (2007a).
12. This book was first published in Danish (Hoffmeyer, 2005). The English version is not just a translation, but a translation by the author together with biosemiotician Donald Favareau, with some new details added.
13. For analysis of his earlier statements, together with a full bibliography of Hoffmeyer to that date, see Emmeche, Kull, & Stjernfelt (2002).

ontological sense, but rather as extremely interesting extensions of a much more general mode of biological organization and interaction that human beings share with all other living creatures." He adds: "So it is namely a question of information, which in one or another sense has a meaning, or is meaningful ... which again assumes a valuable or preferred direction and it is just the kind of information that we associate with *signs* and *sign processes* or *semiosis*" (Hoffmeyer, 2008, p. 12). Thus, life (of organisms) and human mental life belong ontologically together. That is why semiotics cannot be whole without biosemiotics.

Hoffmeyer has repeatedly drawn the attention of biosemioticians to a somewhat paradoxical situation in which the physical approach in biology uses semiotically flavoured terms:

> with such fundamental concepts as *genetic code*, *messenger RNA*, *transcription* and *translation* it is obvious that the revolution of molecular biology, right from its beginnings, implied a powerful semiotic input to biology. ... Molecular biology, the field in which I myself was trained, soon gave rise to a deep understanding of cellular communication processes that indirectly came to pave the way for the formulation of the modern project of *biosemiotics*. (Hoffmeyer, 2008, p. 360, italics in original)

This usage of *semi-semiotic terminology* in molecular biology, which has been called *information talk* (El-Hani, Queiroz, & Emmeche, 2009), has filled the biological textbooks within the last half of century. This has certainly influenced semiotics, via Roman Jakobson's writings *inter alia*. Information talk has induced biologists to think about the problem of roots of information processes, but the theoretical basis for its solution came from elsewhere – from semiotics proper, together with an impact from certain parts of theoretical (epigenetic) biology.

The identification of biosemiotics as a Σ-science can be illustrated by a couple of central emphases made by Hoffmeyer. One of these is the concept of semiotic freedom – the possibility of living creatures to make choices, or, in a better formulation, "the depth of meaning an individual or species is capable of communicating" (Hoffmeyer 2005, p. 434). Another is the importance of including the first-person perspective:

> biology is in principle prevented from including "I" phenomena into its theory structure ... Biology exclusively deals with phenomena that may be described in the language of third-person phenomena, and thus ... excludes this science from arriving at a theoretical understanding of the human biosystem as a first-person being. (Hoffmeyer, 2008, p. 333–334)

He adds, "Biosemiotics must take cognizance of the fact that dynamic systems theory does not exhibit any apparent curiosity about the evolutionary problem of deriving first-person experiential worlds from an ancestry that exhibit nothing but third-person phenomena" (Hoffmeyer, 2008, p. 338).

Biosemiotics as a study of the living world, while taking into account that which is essential for life is that life is the process of knowing, puts quite strong demands on any biosemiotic work. Thus, a rigorous study in biosemiotics should be simultaneously rigorous in terms of semiotics, of theoretical biology, and in terms of empirical biology. Hoffmeyer's work clearly meets these standards.

Empirical biosemiotics is particularly tricky because there is no well-framed and established tradition for doing it.[14] Biosemiotic research has to be based on qualitative methods, but until now the clear understanding of differences between the quantitative and qualitative research methods in biology has not been common. Meaning is not a molecule, but a relation. Accordingly, empirical biosemiotics is a study of relations, functions, distinctions that organisms make, communication, plurality of meaning, and so forth. This is a demonstration of how the sign processes build up the entire diversity of life, how the development of organic forms and their functional organization is related to the grammar of codes and plurality of meanings, what is the role of ontological ambiguity in the formation of organic wholes and ecological balance, how the development and evolution of living systems can be fundamentally understood in terms of transformation between the types of semiosis, how the modelling of living systems has to reflect the modelling processes of life itself, and so forth.

Brief conclusion.

Together with the origin of life and the origin of semiosis, sign relations as the relations of knowing have been produced. Semiotics as the science of knowing, i.e. knowing of knowing, cannot be a whole without biosemiotics, which studies the knowing as it occurs in all forms of life. These formulations depend on our assumption which states that it is founded to generalize the concept of knowing as being applicable for all sign processes.

Biosemiotics studies what life knows. While paradoxically, "it is not knowing, but the love of learning, that characterizes the scientific man" (CP 1.44).

Acknowledgements.

I thank Jesper Hoffmeyer, Paul Cobley, Søren Brier, Donald Favareau, and Terrence Deacon for fruitful discussions that have been very valuable for this work. (A technical note on grants: the European Regional Development Fund, Center of Excellence CECT, SF0182748s06, EMP54, ETF6669.)

References

Anderson, M., & Merrell, F. (1991). Grounding figures and figuring grounds in semiotic modeling. In M. Anderson & F. Merrell (Eds.), *Approaches to Semiotics: Vol. 97. On semiotic modeling* (pp. 3–16). Berlin: Mouton de Gruyter.
Canguilhem, G. (2008). *Knowledge of life*. (S. Geroulanos & D. Ginsburg, Trans.). New York: Fordham University Press. (originally published in 1965)
C.P. See: Peirce, C. S. (1931–1935, & 1958).

---

14. See Kull, Emmeche, & Favareau, 2008.

Deely, J. (2007). *Intentionality and semiotics: A story of mutual fecundation.* Scranton, PA: University of Scranton Press.

El-Hani, C. N., Queiroz, J., & Emmeche, C. (2009). *Tartu Semiotics Library: Vol. 8. Genes, information, and semiosis.* Tartu: Tartu University Press.

Emmeche C., Kull, K., & Stjernfelt, F. (2002). *Tartu Semiotics Library: Vol. 3. Reading Hoffmeyer, rethinking biology.* Tartu: Tartu University Press.

Frost, H. M. (1987). Bone "mass" and the "mechanostat": A proposal. *The Anatomical Record, 219* (1), 1–9.

Hoffmeyer, J. (1996). *Signs of meaning in the universe.* Bloomington: Indiana University Press.

Hoffmeyer, J. (2005). *Biosemiotik: En afhandling om livets tegn og tegnenes liv.* Charlottenlund: Ries.

Hoffmeyer, J. (2008). *Biosemiotics: An examination into the signs of life and the life of signs.* Scranton, PA: Scranton University Press.

Kauffman, L. H. (2005). Virtual logic – the one and the many. *Cybernetics & Human Knowing, 12* (1/2), 159–167.

Krampen, M. (1997). Models of semiosis. In R. Posner, K. Robering, & T. A. Sebeok (Eds.), *Semiotics: A handbook on the sign-theoretic foundations of nature and culture* (vol. 1., pp. 247–287). Berlin: Walter de Gruyter.

Kull, K. (1998). Organism as a self-reading text: anticipation and semiosis. *International Journal of Computing Anticipatory Systems, 1,* 93–104.

Kull, K. (2007a). Biosemiotics and biophysics – the fundamental approaches to the study of life. In M. Barbieri (Ed.), *Introduction to biosemiotics: The new biological synthesis* (pp. 167–177). Berlin: Springer.

Kull, Kalevi (2007b). Life is many: On the methods of biosemiotics. In G. Witzany (Ed.), *Biosemiotics in transdisciplinary contexts: Proceedings of the Gathering in Biosemiotics 6, Salzburg 2006* (pp. 193–202). Salzburg: Umweb.

Kull, K., Emmeche, C., & Favareau, D. (2008). Biosemiotic questions. *Biosemiotics, 1* (1), 41–55.

Levchenko, J., & Salupere, S. (1999). *Tartu Semiotic Library: Vol. 2. Conceptual dictionary of the Tartu-Moscow Semiotic School.* Tartu: Tartu University Press.

Locke, J. (2008). *An essay concerning human understanding* (P. Phemister, Ed.) Oxford: Oxford University Press. (Originally published in 1690)

Lotman, J. (1967). Tezisy k probleme "iskusstvo v ryadu modeliruyuschih sistem." *Trudy po znakovym sistemam (Sign Systems Studies), 3,* 130–145.

Manning, P. K. (1987). *Qualitative Research Methods: Vol. 7. Semiotics and fieldwork.* Newbury Park, CA: Sage Publications.

Merrell, F. (2007). Toward a concept of pluralistic, inter-relational semiosis. *Sign Systems Studies, 35* (1/2), 9–70.

Peirce, C. S. (1931–1935, & 1958). *The collected papers of Charles Sanders Peirce.* Vols. I–VI [C. Hartshorne & P. Weiss, Eds., 1931–1935], Vols. VII–VIII [A. W. Burks, Ed., 1958]. Cambridge, MA: Harvard University Press. (Citations use the common form: CP vol.paragraph).

Rosen, R. (1991). *Life itself: A comprehensive inquiry into the nature, origin, and fabrication of life.* New York: Columbia University Press.

Rosen, R., Pattee, H. H., & Somorjai, R. L. (1979). A symposium in theoretical biology. In P. Buckley & F. D. Peat (Eds.), *A question of physics: Conversations in physics and biology* (pp.84–123). Toronto: University of Toronto Press.

Sebeok, T. A. (2001). *Signs: An introduction to semiotics* (2nd ed.). Toronto: University of Toronto Press.

Short, T. L. (2007). *Peirce's theory of signs.* Cambridge: Cambridge University Press.

Vehkavaara, T. (1998). Extended concept of knowledge for evolutionary epistemology and for biosemiotics: Hierarchies of storage and subject of knowledge. In G. L. Farré, & T. Oksala (Eds.), *Emergence, complexity, hierarchy, organization* (pp. 207–216). Espoo: Finnish Academy of Technology.

Vihalemm, R. (2007). Philosophy of chemistry and the image of science. *Foundations of Science, 12* (3), 223–234.

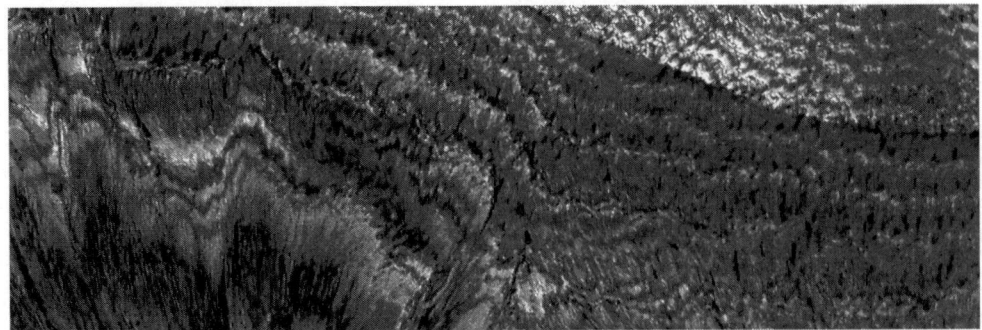

Craft, D. (2009). *Sunrise on Mercury* (detail). Microphoto: Ascorbic Acid Melt Crystals; 34 x 21 cm.

Cybernetics and Human Knowing. Vol. 16, nos. 3-4, pp. 89-106

Musement, Play, Creativity:
Nature's Way

floyd merrell[1]

Charles S. Peirce's notion of *musement* becomes the backdrop for *semiosis*, which is herein considered in terms of nonlinear, nonbivalent, *plurimorphic* processes. These processes not only bear resemblance to certain contemporary scientific theories; they also find expression in Buddhist philosopher Nāgārjuna's middle way. This middle way in turn finds compatibility with Peirce's concept of the sign, which bears on the importance of *play* and *creativity*, as the natural consequence of musement.
Key Terms: Complementarity, Complexity, Emptiness, Entanglement, Musement, Plurimorphity, Process, Tetralemma.

This essay is an inquiry into signs becoming signs, from *possible* signs to actual signs in the physical world or in the world of mind, a *process* which is profoundly germane to Peirce's philosophy. The problem inherent in any effort to articulate the idea of process calls for unorthodox means of description, which can to an extent be found in: (1) alternate non-bivalent logics, and (2) paradoxes of One and Many and Zero and Infinity (as remarkably exemplified in certain aspects of Buddhist thought, especially that of second century philosopher, Nāgārjuna). Such unorthodox means used to account for the process of signs becoming signs, by their very nature, embrace Peirce's seminal thoughts on *musement*, or free-wheeling purposelessly purposeful attunement to emergent signs, which patterns the importance of Peirce's *three categories* of the physical world, of signs, and of thought.

1. Just doing what comes naturally

Given nature's subtle ways, we do not and cannot know precisely what will happen tomorrow, or for that matter, even in the next moment. The unexpected awaits us at every turn, rendering the future a vast, unexplored temporal-spatial expansion. It perpetually recreates the mystery of learned ignorance, as it throws us into the present. When the blinders of our unknowing fall and we are in the present moment of awareness, that's just what we have: the moment, the whole moment, and nothing but the moment.

A *playful mood* allows for, and is usually able to cope with, what might happen to happen. It is free, spontaneous, improvising, and inventive; in short, it is *creative*. In its most creative moments, what happens, happens, as if it were beyond our conscious and conscientious control. Peirce called these moments the play of musement. In his words:

---

1. Purdue University. Email:fmerrell@purdue.edu

There is a certain agreeable occupation of mind which, from its having no distinctive name, I infer is not as commonly practiced as it deserves to be; for indulged in moderately ... it is refreshing enough more than to repay the expenditure. Because it involves no purpose save that of casting aside all serious purpose, I have sometimes been half-inclined to call it reverie with some qualification; but for a frame of mind so antipodal to vacancy and dreaminess such a designation would be too excruciating a misfit. In fact, it is Pure Play. Now, Play, we all know, is a lively exercise of one's powers. Pure Play has no rules, except this very law of liberty. It bloweth where it listeth. It has no purpose, unless recreation .... I will call it "musement" on the whole ... If one who had determined to make trial of Musement as a favorite recreation were to ask me for advice, I should reply as follows: The dawn and the gloaming most invite one to Musement;... It begins passively enough with drinking in the impression ... But impression soon passes into attentive observation, observation into musing, musing into a lively give and take of communion between self and self. If one's observations and reflections are allowed to specialize themselves too much, the Play will be converted into scientific study; and that cannot be pursued in odd half hours. (CP 6.458-59; see also, in general, Sebeok, 1981)

Musement challenges some of our most basic assumptions. It tells us that our intuitions, feelings, sensations, and spontaneity, and our delight in the ephemeral sensations of the moment, our indifference to prestige, riches and power, are as natural as can be for the muser. Thus Peirce counsels us:

I should say, "Enter your skiff of Musement, push off into the lake of thought, and leave the breath of heaven to swell your sail. With your eyes open, awake to what is about or within you, and open conversation with yourself; for such is all meditation." It is, however, not a conversation in words alone, but is illustrated, like a lecture, with diagrams and with experiments. (CP 6.461)

For sure, those who have a penchant for musement in this day and age are swimming against the current. Even so, recently there have been serious studies in psychology, philosophy, anthropology, sociology and literary and cultural studies on musement, and creativity, solely for the sake of contemplating oneself and one's place in the world.[2]

Musement is a state of indifference, with no particular purpose or end. It is a moment of purposeful purposelessness, mindless awareness, passive indeterminacy, all-embracing nothingness. It is in the words of Blaise Pascal, suspension between zero and infinity. For Peirce it is the lively exercise of detached contemplation, when there is neither affirmation nor denial, and at the same time there is both affirmation and denial. There is everything and there is nothing; there is neither choice nor non-choice, only floating dreaminess.

---

2. To cite a few representative studies that have a bearing on this essay, Abram (1996), Borgo (2005), Carse (1986), Combs (2000), Csikszentmihalyi (1997, 2007), Fauconnier and Turner (2002), Gardner (1993), Kaufman and Baer (2006), Michalko (2001), Miller, D. L. (1989), Miller, A. (2002), Nachmanovitch (1990), Pope (2005), Sawyer (2006), Shlain (1993), Rothenberg and Hausman (1976), Root-Bernstein (1999), Weisberg (2006), West (1997, 2004).

2. Musement's intrigue

A musing act is carried out for its own sake. It is self-contained, self-sufficient, and self-reflexive. It cannot be controlled or manipulated in order to gain something that is other than what it, itself, *is*. It is not a step toward something else; it has no further intentionality. It demands a sort of naïve innocence. In this sense, the act of musement is a completely free act, unconditioned except by its own impulse. In a word, musement simply is; *it is what it is.*

But jumping to the conclusion that the formula, It is what it is, entails hard-rock identity is like the dog barking up the wrong tree. What the animal honing his hunting skills thinks there is up there, there *is not*. What is, is what it is only in *inter-dependent inter-relationship* with what it is not. However, like pure play or musement, there's no telling what the future holds, so what is not in the next moment can *become something other than what it was becoming* (hereafter BSO) to ephemerally take on the countenance of what is, and then in the next moment it will have passed into what it *was not*. And so on.

Obviously, musement occurs in the *now* moment. The muser slows this moment, almost to a standstill. Even when past and future are contemplated, interest remains absorbed in the present. Each moment is full in and of itself, and then in the blink of an eye it begets the next moment. Each moment is a crossing that contains the past because, having expired, it has emerged, and it contains the future because, although that future hasn't yet dawned, it is there, in the now, within the field of all *possibilities*.

What are the implications of the muser's now? Above all, she has a totally different attitude toward work. She does not merely work; she strives to get in tune with *working*, as act, action, or better, as *acting process*. Work is for her an ongoing process, a *doing* for the sake of doing rather than to be done with it, pick up the check or reap the rewards, and ultimately gain power, fame and fortune, and realize success. If the doing isn't meaningful in itself, she simply won't engage in it. Her respect for her doing is characteristic of the musing attitude. She cultivates a plant, for example, but cultivating it is meaningful in itself; it is ennobling regarding the very process of *acting*, which is like *playing*, or *play acting*. While creating complementariness between herself and the plant-becoming, she is in the process of herself-becoming; she is becoming nature and nature is becoming her. This play acting is nature's way in the most profound sense. The muser's play acting coalesces with her working, consequently she has little regard for the product of her work. She enters into the process for the sake of that process itself, and not for what might come of it. Her play acting is comparable to the dancer in the process of dance-becoming which is in the process of dancer-becoming, or comparable to the musician's instrument-becoming and the instrument's and musician's jazz-becoming at the same time that the jazz is musician-becoming and instrument-becoming. She forgets herself, loses herself in her play acting: She is selflessness-becoming. This is not to say that there is no consciousness. The muser is conscious, for sure, but her consciousness is processual: she is in the process of *selfing*.

In sum, the muser's activity is neither selfish nor utilitarian nor the product of mindless rule following. She acts under neither compulsion nor obligation. She is driven neither by the desire for immediate rewards nor for power and notoriety. So what motivates her? I would suggest that it is a desire, from deep within, to go where she's never before gone, where perhaps few before her have gone (see Nachmanovitch, 1990).

3. When There's Nothing (That Is, *Emptiness*) Instead of Something

The musing process is above all a matter of Peirce's *Firstness-becoming*. Firstness, of Peirce's three categories of the physical world, signs, and thought—or mind—is (1) what it is *becoming*, irrespective of (2) anything *other* than what it is becoming (*Secondness-becoming*), before there is any form or fashion of (3) *Thirdness-becoming*, whose function entails the *becoming of the process of mediating* Firstness-becoming and Secondness-becoming in the same way it brings about a *mediation process* between itself and them (CP 1.300-53).[3]

There is not (yet) any otherness (Secondness) with respect to Firstness, and hence there cannot (yet) be any Thirdness providing a mediating process. Genuine *semiosis* entails the three categories operating in concert. United, the process flows along; divided, it disintegrates. In this sense, the categories are *inter-dependent, inter-related,* and *inter-active* (hereafter i-i-i-), and, as we shall note, in their composite form their process-of-becoming is *contradictorily complementarily coalescent* (hereafter CCC) (CP 1.300-21, CP 2.233-42, CP 6.272-86).[4] Musing, chiefly of the nature of Firstness, is subtle, supple, rich, and pregnant with possibilities; Secondness is the necessary otherness of Firstness, integrating it into the mind's and the world's becoming. Thirdness is the swirling, whirling flow that brings them into coalescence at the same time that it coalesces with them. Then there's language, premiere Thirdness, proud, presumably objective language, which, using proper logic and reason, punctures, mutilates and ruptures the flow in order to taxonomize all that is becoming. But necessary language, for how, otherwise, could the *semiosic flow* become itemized and articulated, however inadequately?

And yet, semiosis flows on. As process, the inter-dependent, inter-related, and inter-active (i-i-i-) and the contradictorily complementarily coalescent (CCC) nature of semiosis patterns (1) *complexity* and *chaos* in the physical sciences, especially in

---

3. For discussion of the categories, see Almeder (1980), Esposito (1980), Hookway (1985), Stearns (1952). In this essay I wish to qualify the categories not as fixed but always in formation, not stable fact but flowing within changing contexts. They are always becoming something other than what they are becoming (BSO), somewhat in the order of Deleuze and Guattari's (1983) rhetorical display upon accounting for the deterritorializing process, and especially following quantum theorist John Archibald Wheeler's (1980a, 1980b, 1984) notion of the universe's process.

4. As we shall note, the process can hold contradictions together in its liquid embrace (contrary to bivalent Contradiction barring); it is complementary, since there is neither exclusively the one horn of a contradiction nor the other, but something else, something new emerging; and it is coalescent, for whatever *is*, is always becoming something other (see Costa, 1974; Costa & Krause, 2001, 2003; Priest, 1987, 2004, 2006; Rescher & Brandom 1979, and application in merrell, 1998, 2007).

Ilya Prigogine's conception, and (2) *entanglement*, implicit most dramatically in quantum theory, a process that radically entails the inter-connectivity of everything.

Prigogine's inorganic dissipative structures manifest organic, biological, life-like principles. Dissipative structures are open systems, exchanging energy and matter with their environment. As these systems become increasingly complex, they enter into far-from-equilibrium conditions that eventually reach a threshold after which the systems dissipate and spontaneously become self-organizing. Prigogine self-organizing systems are germane to the semiosic notions of i-i-i- (everything is inter-connected), CCC (everything is in the process of fusing into everything else, becoming increasingly complex), and BSO (everything is always in the process of becoming some other). Since these processes evolve into dynamic life-like systems, local perturbations can bring about global changes that become radically entangled. The concept of entanglement, from the quantum world to galaxies, also tells us that everything is i-i-i-, CCC, and BSO, such that whenever something happens to one phenomenon it has an effect on all phenomena. While this effect in the quantum micro-world is instantaneous, in our concrete macro-world, time plays a role; hence mutual inter-action of the categories and of signs, their otherness and their mediation, is always to a lesser or greater extent delayed (Wheeler, 1980a, 1980b).[5]

Entangled complexity regarding world processes comes about through access to the deeper environs of musing, within one's own reality, the social reality of one's community, and one's physical reality, [none of these terms seems to be used ironically] which can only be fathomed by wordless feeling, emoting, intuiting, imagining, and sensing. Then, and only then, after musing moments, can those moments be put into language, however loose and vague. The problem is that language, whatever language, whether logic, mathematics, Boolean computer formalism, or natural language, is what it is only with respect to what it is not: silence, the blank page, zero, or in the considerably more profound Mahayana Buddhist sense, emptiness. In other words, musing moments are pre-language, pre-Firstness, pre-semiotic (Baer, 1988) (CP 6.185-222).[6]

4. Patterning the process

How might it be possible to account for this emptiness with respect to Peirce's concept of the sign? Consider figure 1.

5. From diverse views, on the becoming of time with respect to semiosis and the categories, see Hartshorne (1970), Kauffman (2002b), Rosenthal (2000); on dissipative structures and complexity, see Prigogine (1980, 1997), Prigogine and Stengers (1983), Nicolis and Prigogine (1989); on entanglement and fractals, see Aczel (2003), Clegg (2006), Mandelbrot (2004), Pagels (1988).

6. I should at this juncture point out that Søren Brier brought to my attention his seminal article on Peirce's panentheistic scientific mysticism (2008) wherein he carefully documents Peirce's own notion of emptiness. This essay, I would hope, can serve as a companion to Brier's work.

floyd merrell

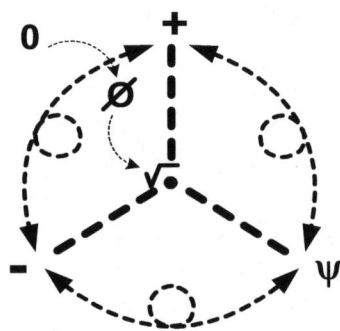

Figure 1

Notice that three-way diagram suggests no more than the mere possible possibility of a sign: there is actually no sign, its other, or mediation by means of which meaning might be forthcoming. There is just possible possibility. Thus the lines of the diagram are dotted, since nothing is (yet) actualized. The swirling, swiveling, rippling, scintillating process is: $0 \approx \emptyset \approx \sqrt{\bullet} \approx \blacktriangle (= + \approx - \approx \Psi)$ (where $\approx$ depicts multiple possible nonlinear transitory paths). The symbols making up the equation for this process can be qualified as:

(1)   0, zero, nothingness or emptiness in the Buddhist sense, the range of all *possible possibilities*, [7]

(2) $\emptyset$, the Empty Set, or noticed absence of the possibility that some sign or set of signs could be there but it is not, or it was there and now it is not, but it *might be* there once again at some future moment,

(3) $\sqrt{\bullet}$, the beginning of that necessary demarcation specifying what possibly is (Firstness) and what is possibly other than what possibly is (Secondness) (+, –),

(4) $\blacktriangle$, representing the three lines—or the possibility of Peirce's categories—making up the naturally democratic tripodal figure as a model of the sign,

(5) Ψ, bringing about mediation (Thirdness) of + (Firstness) and – (Second-ness) in the same that it mediates between itself and them, democratically.

In the beginning there is no more than emptiness (0), the possible possibility of a sign; then there is the noticed absence ($\emptyset$), of what might have been or never was but

7.   The Greeks feared both zero and infinity. Zero threatened to give itself membership into the collection of numbers as something which was nothing, while infinity masqueraded as an uncountable number of things that, like zero, couldn't be specified (Barrow, 2005, pp. 23-26).

might possibly be, emerging; then √• (a strange counterpart to the irresolvable imaginary number, √-1) arises, and it can evoke the possible presence of a sign (+), and the absence of that with which it might possibly interact (–), an other; and finally, the possibility of mediation emerges (Ψ). The psi symbol suggests mediation bringing + and – together and at the same time bringing itself into i-i-i- with them in such a manner that a possible interpretation, or meaning as it were, may begin its process of emerging. From within this range of possible possibilities, nature's way is germane to the semiosic process, and our imperious, language-driven self-conscious cognizing can hardly do more than go along for the ride.

This is anti-essentialism in the most radical sense. That is, i-i-i- is a matter of radically CCC oriented process. There is neither substance nor product, for i-i-i- and CCC involve a flow of signs. There is no Cartesian clarity and distinction regarding this process. Thoughts and words are, and will remain, tinged with a greater or lesser degree of vagueness, ambiguity, incoherence, and paradox—paradox, that necessary ingredient in the most fruitful mixes of musement. In spite of our desire to hammer thoughts and words into well-honed instruments, the semiosic process is always around to reveal our shortsightedness, our limitations, our learned ignorance. And yet, we are condemned to that prison-house of words. It is nature's way with respect to the human animal. So, we speak and we write and get along as best we can with the signs we have at hand.

It is becoming increasingly apparent in our times that when saying and writing are at their best, they are never monological or monistic, nor are they simply dualistic; rather, they are dialogical, in the triadic sense, and such dialogism, when at its best, remains attuned to nature's way, from figure 1 to actualized and properly interpreted signs. However, it seems that there is always that ubiquitous dualism: silence/word. But silence is not mere absence; it is the mute possibility of all possible words. And words are not the mere absence of silence, which would entail the makings of figure 1; yet they are inter-dependent with silence, since silence bears the possible possibility of any and all words that can emerge to accompany the words that in some past moment broke the silence to emerge as the words they are at the present moment. In sum, the relationship between silence and words is nondualistic and nonmonistic; it is a matter of triadic complementarity, in the playful, creative, musing now.

This inevitably introduces us to that disconcerting process within which we flow …

5. From Nothing to One to Many: *Plurimorphity*

Human cultures are guided by their particular form of a collective imagination—through moments of musement—that can take on diverse countenances from one culture to another. The very idea of collective imagination suggests a repository of premonitions, presuppositions, prejudices and proclivities that have accumulated in past times among human communities, and it suggests development of expectations with respect to what will emerge in future times. The future holds surprises in store, in large part because both the past and the present are always becoming something other than what they were becoming (BSO). This process evokes the notion of pluralism.[8]

However pluralism may be interpreted in terms of an individual within a particular human community, it is usually in various stages of transition into something other than what it was according to the following: (1) Pluralism embodies distinctions: differences that make a difference, (2) Yet, pluralism—difference s because the community opinion says so—can fade into differences that are simply different; consequently, the differences become increasingly diffuse, hence (3) the process of diffusion begins spilling into *plurimorphity*, that is, contradictory, complementary convergent heterogeneities, differences that make new and different differences.[9]

This much tentatively outlined, the paradox of the One and the Many emerges. However, there may be a way out of this not-so-malicious-paradox in the name of what has been termed…

5a. Plurimorphity as One and Many, and Neither One Nor Many

I prefer to qualify the radically fluid, fluctuating semiosic process as plurimorphity, since within plurimorphity there is no collection of relatively autonomous entities at war with one another, as bivalent thinking might have it. There are not simply distinctions and differences, for anything and everything is perpetually emerging and changing (it is i-i-i- and CCC). This is a qualitative notion that cannot be determinately and uniquely quantified, at least by the use of linear formal language, because diversity within plurimorphity is radically nonlinear, as in Prigogine complexity. The concept of pluralism, unfortunately, is often construed as a matter of bivalent diversity. Nonlinear, *n*-valent plurimorphity, in contrast, is emergent novelty; it is fresh, spontaneous, and creative, available to the musing mind.

Moreover, BSO in the CCC sense finds itself unfit for convenient inclusion within what currently goes as the linguistic turn, that is, insofar as language is conceived in linear fashion.[10] However, BSO, as well as CCC and i-i-i-, mesh harmoniously with radically extralinguistic Buddhist Tetralemma, for which second century philosopher, Nāgārjuna, is notorious. According to the Tetralemma, what is under consideration, if it is a particular something that can be taken for what it is, is a matter of its possible qualification from multiply divergent possible perspectives. These possible perspectives are qualified according to the following injunctions regarding whatever

---

8. For work bearing on Peirce's pluralistic leanings, Anderson (1987), Dozoretz (1979), Esposito (1980), Rescher (1993), Rosenthal (1994). *Pluralism*, as the concept unfolds here, will give way to a more processual concept, *plurimorphity*. The sources for these two terms with respect to this chapter are multiple, from diverse perspectives and disciplines (see Beall & Restall 2006; Connolly, 1995; Plaw, 2005).

9. It bears mentioning that *plurimorphity* does not raise the question of compatibility or incompatibility in regard to *heterogeneity*, because the *unity* of the community within which this *heterogeneity* pervades is already a tacitly accepted fact. The community is a community with comparable possibilities of becoming among all members, but members make different selections from those possibilities according to their own whims and wishes and modes of thinking and believing.

10. This is, of course, an exceedingly broad topic into which I cannot enter here (in general see Bell, 1998; Berman, 1988; Brandom, 2008).

assertion is up for consideration: (1) "It is so!", (2) "It is not so!", (3) "It is *both*!", and (4) "It is *neither*!". And then, to do proper justice to the Tetralemma, Nāgārjuna adds what might be dubbed a couple of corollaries: (5) "All of the above," and (6) "None of the above."

"Outlandish!", one might immediately with to blurt out. Granted, illogical the Tetralemma certainly appears, and we naturally tend to shrink back when confronting such apparent illogic. However, if we integrate the Tetralemma with our standard notions of Zero and Infinity, and One and Many, the Tetralemma becomes less unwieldy.

The classical logical Principles of Identity, Non-Contradiction and Excluded-Middle allow for Truth, Falsity, and what logical positivists often classified as Nonsense or Meaninglessness. So, to a truth-claim, the possible answers are Yes! or No!, True! or False! Nāgārjuna adds, apparently *contradictorily* and *inconsistently*, Both!, which reminds one of Niels Bohr's Complementarity Principle allowing, intermittently, for a subatomic phenomenon as now a wave amplitude, now a particle, but not both within the same timespace context. And Nāgārjuna further declares Neither! Neither, because there might be some third possibility emerging from between the either and the or, in a manner of speaking. To top it all off, Nāgārjuna then includes "All of the above!" How so? Because, before any possibility emerged into the light of day, it was included within the range of all possible possibilities, that is, within emptiness. And Nāgārjuna also includes "None of the above!" since no matter how many possibilities have begun their process of becoming, there is always something else possible, for whatever is in the process of *BSO*, it has remained, and will always remain, incomplete.[11]

If we integrate the Tetralemma and its corollaries with our standard notions of Zero and Infinity, One and Many, it becomes somewhat less unwieldy. Zero and Infinity—and by extension One and Many—are the Twiddledee and Twiddledum of numbers, and indeed, of thought in general. In many respects they are the mirror image of one another. Multiply zero by any number and you get zero; multiply Infinity by any number and you get Infinity. The same equals the same. Dividing a number by zero leaves you with an undefined answer, tantamount to Infinity; dividing a number by Infinity leaves you with zero. Add zero to any number and the number remains unchanged; add Infinity to any number and the yield is Infinity, unchanged. Zero and Infinity are two sides of the same coin, two sides of a sheet of paper. They are the *Yin* and *Yang* of all that is becoming; they are the beginning and the end of the process of becoming that has neither beginning nor ending. They are what is and what is not, positive and negative, + and − , of figure 1. They are the two poles of becoming by means of which what is (+) is always becoming something other (−) than what it was

---

11. For further I would suggest consultation of Gangadean (1981), Hall and Ames (2001), Huntington (1989), Kalupahana (1986), Laycock (1994), Loy (1989), McCagney (1997), Mansfield (1989), Odin (1996), and especially to a translation of Nāgārjuna's "Fundamentals of the Middle Way" in F. J. Streng's *Emptiness: A Study of Religious Meaning* (1967).

becoming (BSO), but that *other than* is what is also already becoming something other than what it was becoming (Seife, 2000, pp. 19-23, 106-13, 131-32).

However, there seems to be something lacking in an equation containing only + and − . There must be some middle ground, some *middle way*, some moderator whose mediation not only takes in + and − , but, in addition, takes in itself in its mediation of itself in regard to + and − . The mediator, or ψ in figure 1, as it were, mediates what is *other than* what it *is*, and at the same time it mediates itself, reflexively. In this manner, if + is what is in the BSO process, and if − is what + is not in the process, then ψ entails both + and − , and at the same time it is neither + nor − . This, in what might appear as a convoluted—and perhaps muddled—frame of reference, is the Tetralemma in another form: (1) +; (2) − ; (3) both + and − ; and (4) neither + nor − , "All above" (… ∞), and "None of the above" (… 0) (where "All of the above" implies the infinite march toward completion, and "None of the above" implies the spontaneous emergence from emptiness) (see also merrell, 1998, in press).

The Tetralemma throws *Truth* as One for a loop, and *truth* as Many enters the fray and plays loosely, while allowing for the truth of particular traditions, which can be properly understood only within the very traditions that have elaborated them. Yet each tradition, qualified in terms of i-i-i- and BSO, enjoys some commonalities with other traditions: they are, in their composite, CCC, that is, complementarily plurimorphic, within a complexly entangled system. In other words, given its radically nonlinear nature, figure 1, the possible possibility of signs becoming signs, is always possibly moving out along multiple diverging and converging paths in a diversity of possible directions, breeding plurimorphy. The Tetralemma entails perspectivism of the most radical sort.

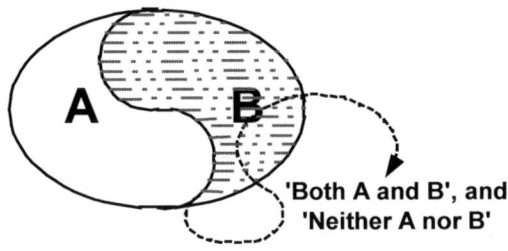

'Both A and B', and
'Neither A nor B'

Figure 2

A simple diagram can help illustrate this notion. Figure 2 depicts two values, A and B, with a line of demarcation separating them. According to the tenets of classical logical principles, A is what it is and it cannot be anything else, and the same can be said of B. But what is the line? Is it A? No. B? Negative also. It is A-less and B-less. But as A-less, it shares some value with B: It is A-less. And as B-less, it shares some value with A: It is B-less. In a manner of speaking, then, the line is Both A and B, and in this respect it can find a comfortable resting place within the sphere of Firstness, as the possibility of signs, some of which are contradictory with other sign possibilities. But that's no problem, since one compound symbolic sign, say, the *Earth as center of the Universe*, can rest quite comfortably with another compound symbolic sign, the *Sun as center of the Universe*, as long as they are no more than possibly possible signs. Thus the classical principle of non-contradiction doesn't necessarily hold for mere possible possibilities.

Once signs are in the process of becoming actualized, they begin by taking their place among other signs. Consequently, inconsistencies occasionally raise their ugly heads. One sign, *Sun, now* considered the center of the Universe, enters into conflict with another sign, *Earth*, which was center of the Universe. Some new interpretation emerged, as if from the line in figure 2. This new sign, *Sun*, is what the old sign, *Earth*, was—center—and it is not what that old sign was—it is *Sun*. It is neither what the old sign was nor what it was not, for another possible sign has emerged to take its place as the center of the Universe. Thus the classical principle of excluded-middle doesn't necessarily hold, at least when we include different complex, entangled timespace contexts in the picture.

This reveals that signs as *generalities* are invariably vague and incomplete (CP 1.463-69, CP 5.506-16, CP 5.447-57), and that somewhere along the winding path toward more adequate generality, alternatives (Neither the one nor the other) stand a chance of emerging, within a new timespace context, from within the included-middle, the middle way, that would otherwise have been the excluded-middle principle. The plurimorphic process, emerging from the middle way—metaphorically, like the line in figure 2—is nature's way. Classical logic holds fast to the excluded-middle principle; but according to some alternative logics, the middle way, or an included-middle principle—fit for everyday concrete practice—makes its entry.[12]

5b. A brief digression

Figure 2, as we shall note more explicitly below, is commensurate with Nāgārjuna's thought as grounded in the Tetralemma. Some of Nāgārjuna's critics say he destroyed logic in order to bring out his interpretation of the Buddhist notion of emptiness. Other critics contend that Nāgārjuna's logic is evasive, for he asserts, simultaneously, the equivalent of "Either to be or not to be," "Both to be and not to be," and "Neither to be nor not to be."

In response to these charges, Shotura Iida (1980) offers the following scheme (see figure 3). Nothing in this figure reveals emptiness, or in other words, zero. Yet it is there. It is like the Buddhist wheel. For the wheel to revolve, the center must be Empty; it must be of zero rotation; thus it is tantamount to $0 = 0$. There is no value, no rotation either in the one direction (+) or in the other direction (−). Likewise, for numbers to exist, for there to be a distinction, for there to be something rather than

---

12. From various angles regarding alternative logics, see Costa (1974), Costa and Krause (2001, 2003), Heelan (1971, 1973, 1983), Priest (1987, 2004), Priest, Routley, and Norman (1989), Putnam (1971, 1983).

nothing – this or that, this and that, neither this nor that, neither existence nor nonexistence – there must be emptiness, zero. It's as simple and as mind-boggling complex as that

.

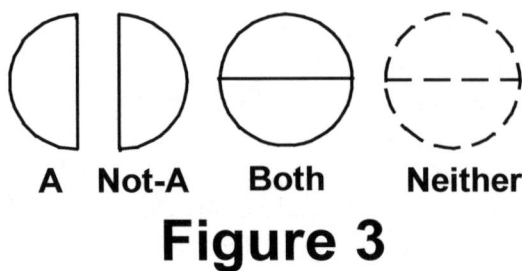

A   Not-A   Both   Neither

Figure 3

The *both-and* and *neither-nor* make up Nāgārjuna's middle way is patterned in the line of demarcation in Figure 2 that offers the possibility of something new emerging. The dividing line is like the liar paradox, "I am lying"—presumably the prototypical example of a Gödel number in natural language—which is both true and false, and at the same time it is neither true nor false (Goldstein, 2005). And yet, this something else is of some nature outside the ordinary confines of the lines making up the diagram. In a manner of speaking, it is of the nature of pure possible possibilities (or emptiness) offering themselves up as candidates for beginning their becoming into the light of day, into what might be taken as some aspect of our complexly entangled real world.[13]

6. Of the Middle Way, Again

But a question arises: How does *plurimorphity* emerge? In a nutshell, from One, Two emerges, then Three, and then Many (as in Lao-Tsu, 1963). It is Yin-Yang rhythm. It is a matter of imbalance, disharmony, decoherence, dissonance, disequilibrium, and syncopation, and above all, syncopation. Three gives us "A One … a Two …, … and a Three," syncopatedly speaking. Three creates, like syncopation, asymmetry, hesitation, vacillation, a pause that doesn't refresh, not yet at least, but rather, there is a moment of uncertainty, doubt, undecidability, during musement, the premonition that something will happen, but it is not yet known what.

Then a surprise suddenly erupts, and something different and new emerges. Ah, so that's it! It's the surprise that refreshes. It evokes, it provokes, it pushes toward who

---

13. George Spencer-Brown (1979), beginning with the originary *mark of distinction*, and in view of the Liar Paradox and its mathematical counterpart, $\sqrt{-1}$, describes this process as imaginary oscillating, stuttering, flip-flops between the either and the or, + and − . This is a self-referential, self-informing form that offers no ready-made solution; it presents a quandary, a paradox; it can do no more than vibrate and vacillate; it remains caught up in scintillating, quivering, indecision (for further, see Engstrom, 1999, 2001; Herbst, 1993; Kauffman, 2001, 2002a; Kauffman & Varela, 1980; Robertson, 1999; and Schiltz, 2002).

knows where or when. It creates that moment when many possibilities are there and waiting; then something spontaneous and new suddenly makes its appearance. And we are in the nonlinear, unpredictable, swiveling, swerving, spiraling path where plurimorphity pervades. This path periodically—and ephemerally—opens up the middle way, the nonbivalent included-middle, which allows for possible possibilities rather than categorically barring any and all Contradictions by way of the excluded-middle.

In this manner, plurimorphity is a matter of creativity. It begins with musement, then there's syncopated Threeness, with a feeling of something as yet unspecified and perhaps unspecifiable. Then, fingers do the walking, eyes do the probing and scanning, and ears, nose and tongue do the sensing, when the proprioceptive, kinesthetic, somatic body does the talking, in its silent, nonverbal way: in a manner of speaking, nature's way. During such spontaneous corporeal activity and nonverbal dialogue, mind is not just along for the ride. Mind and body, *bodymind,* as a complementary plurimorphic whole, enters the creative vortex.

Plurimorphic play, creativity, and musement through an act of *abduction,* as a consequence of the emergence of a First, of Firstness.[14] The creative act, in other words, is in the process of bringing about an abducted possibility that is always BSO. But the act does not simply appear, as if out of the clear blue sky. It emerges, because creative people have an insatiable curiosity. They are always on the lookout for the new and different. When confronting a perplexity, their persistence simply won't allow them to leave it and move on to other less confounding pastimes; they are tireless workers, and yet, they can find time for contemplation, pondering, musing, which prepares the terrain for creativity (Anderson, 1987; Hausman, 1975; Rothenberg, 1990).

Albert Einstein's unique thought experiments are certainly among the most salient cases of scientific plurimorphic play, creativity, and musement. After he released his special theory of relativity to the world in 1905, he soon knew there was something missing. Filling this gap took him 11 years, culminating in his general theory of relativity, which addressed problems in Newton's theory of gravity in view of the special theory. While pondering over the apparently irreconcilable differences between Newton's theory and the special theory, it occurred to him that if an observer is in a state of free fall, there exists, from within the frame of reference of that observer, no gravitational field. If while falling, the observer lets go of an object she has in her hand, it won't fall; but from an outside frame of reference both she and the object will be falling according to Newtonian gravitational force. From within her perspective, she is in a state of rest; from the outside perspective, she is in a state of

---

14. The abductive process emerges out of musement. It customarily involves acts of improvisation, above all in the sense that improvisation is the "skill of using bodies, space, all human resources, to generate a coherent physical expression of an idea, a situation, a character …; to do this spontaneously, in response to the immediate stimuli of one's environment, and to do it … as though taken by surprise, without preconceptions" (Frost & Yarrow, 1990, p. 1). For a survey, and arguments pro and con, of abduction, I would suggest Fann (1970), Queiroz and merrell (2005), Turrisi (1990), and Wirth (1999).

free fall and accelerating at the rate of 32 feet per second squared. At the outset, an ordinary mortal would discard the very notion as illogical. For Einstein, the apparent contradiction was eventually resolved. He had the ability to hold disparate concepts together in order to resolve the problem situation their conjunction produced (West, 2004).

The usual tendency is to think of Einstein's solving his physics problems with mathematics. Even psychologist Howard Gardner in *Creating Minds* (1993) describes Einstein as the premiere example of a logico-mathematical mind. However, Gerald Holton (1978), Walter Isaacson (2007), Arthur Miller (2002) and Leonard Shlain (1993) emphasize the fact that Einstein was actually relatively weak in mathematics, occasionally relying on professional mathematicians for the necessary equations to illustrate his ideas. Einstein revealed the unmathematical nature of his mental strength to his psychologist contemporary Jacques Hadamard: "The words of language, as they are written or spoken, do not seem to play any role in my mechanism of thought. The psychical entities which seem to serve as elements in thought are certain signs and more or less clear images which can be 'voluntarily' reproduced and combined.... The above mentioned elements are, in my case, of visual and some of muscular type" (Hadamard' 1945, pp. 142-43).[15]

In short, Einstein's creativity emerged at the kinesthetic-proprioceptive-somatic level, within musement, pure play, outside the customary parameters of logic and reason and thought. Einstein's particular form of creativity, I would suggest, is the focus of this entire essay: it is prelinguistic, prelogical; it is chiefly of the nature of Peirce's Firstness; it precedes perception and conception, Peirce's Secondness and Thirdness; it is barely entering into the process of emerging from the range of all possible possibilities.

7. The centrality of *plurimorphity*

The notion of non-cerebral, non-conscious, kinesthetic-proprioceptive-somatic feeling before sensing and thinking, before labeling and cognizing, without clearly and distinctly being able to say how it is one knows at this level, has always been commonplace for some artists, scientists, and thinkers and writers. It recalls Blaise Pascal's heart that *has reasons that reason cannot know*. It also recalls Picasso, who once told a friend: "I don't know in advance what I am going to put on canvas any more than I decide beforehand what colors I am going to use ... Each time I undertake to paint a picture I have a sensation of leaping into space. I never know whether I shall fall on my feet. It is only later than I begin to estimate more exactly the effect of my work" (in Ashton, 1972, p. 28).

This Firstness, this feeling for what is becoming and what might be thought and said bears directly on plurimorphity. Unlike pluralism, whose focus tends to rest on relatively fixed entities, plurimorphity involves ongoing process. It is a form whose

---

15. In fact, Einstein once confessed to Max Wertheimer that he only vaguely understood the implications of his visual and muscular sensations, for his feelings were 'very hard to express' (Wertheimer, 1959, p. 228, n.7).

content is no content without the form and the form is nothing without the content. It is One, since it can't be subdivided; yet it is Many, because Many involve CCC through i-i-i- in order that they may hopefully become One. Plurimorphity is virtually infinitely pliable—form-able—since it is radically BSO; yet it remains what it is with respect to some other that is also BSO in complementation with it. It is Firstness because it is One, but it engenders Secondness because there is its Other, and mediating Thirdness emerges to bring them together in the same manner in which it unites itself with them.

Indeed, we can wrap this essay up with allusions to Peirce's categories as process:

- First is potentially and concretely creative—and perhaps playful, musing—feeling (of what is such as it is, with no consciousness of anything other than what is: Nature's Firstness, because it is not yet available to our self-conscious, cogitating, abstracting, intellectualizing self).
- Then sensation of what that feeling is about as so-and-so (the act of becoming concretely aware of something other than the subjective self and/or other than the sign: Secondness).
- A split second later, there is awareness *of* what that 'so-and-so' is, because, according to one's penchant for classifying and generalization by means of conventional knowing, and/or one's personal idiosyncratic knowing, it is of such-and-such a set of characteristics (the act of re-cognizing that something and endowing it with meaning: Firstness and Secondness mediated by Thirdness).

Thus, the plurimorphic nature of ourselves and our creative—perhaps playful, musing—imaginary worlds, our diverse community of like but different individuals, and the physical world, are all in CCC through i-i-i-, which is always already BSO. These three abbreviated terms, I would modestly submit, are germane to Peirce's process philosophy, which can give us a novel take on trans- and intercultural, and trans- and interdisciplinary, plurimorphity, and how it is that our physical and cultural worlds of entangled complexity, and ourselves as well since we are signs among signs, are nevertheless capable of rich self-organization.

References

Abram, D. (1996). *The spell of the sensuous: Language and perception in a more-than-human world*. New York: Pantheon.

Aczel, A. D. (2003). *Entanglement*. New York: Plume.

Almeder, R. (1980). *The philosophy of Charles S. Peirce: A critical introduction*. Totowa, NJ: Rowman and Littlefield.

Anderson, D. (1987). *Creativity and the philosophy of C. S. Peirce*. Dordrecht: Martinus Nijhoff.

Ashton, D. (Ed.) (1972). *Picasso on art: A selection of views*. New York: DaCapo.

Baer, E. (1988). *Medical semiotics*. Lanham: University Press of America.

Barrow, J. D. (2005). *The infinite book: A short guide to the boundless, timeless and endless*. New York: Random House.

Beall, J. C., & Restall, G. (2006). *Logical pluralism*. New York: Oxford University Press.

Bell, J. A. (1998). *The Problem of difference: Phenomenology and poststructuralism*. Toronto: University of Toronto Press.

Berman, A. (1988). *From the new criticism to deconstruction: The reception of structuralism and post-structuralism*. Champaign-Urbana, IL: Illinois University Press.

Borgo, D. (2005). *Sync or swarm: Improvising music in a complex age*. New York: Continuum.

Brandom, R. B. (2008). *Between saying and doing: Toward an analytic pragmatism*. New York: Oxford University Press.

Brier, S. (2008). A Peircean panentheistic mysticism. *International Journal of Transpersonal Studies, 27*, 20-45.

Carse, J. P. (1986). *Finite and infinite games: A vision of life as play and possibility*. New York: Free Press.

Clegg, B. (2006). *The god effect: Quantum entanglement, science's strangest phenomenon*. New York: Macmillan.

Combs, J. E. (2000). *Play world: The emergence of the new Ludenic age*. Westport, CT: Praeger.

Connolly, W. E. (1995). *Pluralism*. Durham, NC: Duke University Press.

Costa, N. C. A. da (1974). On the theory of inconsistent formal systems. *Notre Dame Journal of Formal Logic, 15*, 497-510.

Costa, N. C. A. da, & Krause, D. (2001). *Complementarity and paraconsistency*. Retrieved November 17, 2009 from: http://www.cfh.ufsc.br/~dkrause/CosKra00.pdf.

Costa, N. C. A. da, & Krause, D. (2003). *The logic of complementarity*. Available Online (as of November 17, 2009) from: http://philsci-archive.pitt.edu.

Csikszentmihalyi, M. (1997). *Creativity: Flow and the psychology of discovery and invention*. New York: Harper.

Csikszentmihalyi, M. (2007). *Everyday creativity and new views of human nature: Psychological, social, and spiritual perspectives*. Washington, DC: American Psychological Association.

Deleuze, G., & Guattari, F. (1983). *Anti-Oedipus: Capitalism and schizophrenia*. Minneapolis, MN: University of Minnesota Press.

Dozoretz, J. (1979). The internally real, the fictitious, and the indubitable. In K. Ketner & J. Ransdell (Eds.), *Studies in Peirce's Semiotic I*(pp. 77-87). Lubbock, TX: Institute for Studies in Pragmaticism.

Engstrom, J. (1999). G. Spencer-Brown's *laws of form* as a revolutionary, unifying notation. *Semiotica, 125* (1/4), 33-40.

Engstrom, J. (2001). Precursors to *laws of form* in C. S. Peirce's collected papers. *Cybernetics & Human Knowing, 8* (1-2), 25-66.

Esposito, J. (1980). *Evolutionary metaphysics: The development of Peirce's theory of categories*. Athens, OH: Ohio University Press.

Fann, K. T. (1970). *Peirce's theory of abduction*. The Hague: Martinus Nijhoff.

Fauconnier, G., & Turner, M. (2002). *The way we think: Conceptual blending and the mind's hidden complexities*. New York: Basic Books.

Frost, A., & Yarrow, R. (1999). *Improvisation in drama*. New York: Macmillan.

Gangadean, A. K. (1981). Nagarjuna, Aristotle, and Frege on the nature of thought. In N. Katz (Ed.), *Buddhist and Western philosophy* (pp. 202-43). Atlantic Highlands, NJ: Humanities Press.

Gardner, H. (1993). *Creating minds: An anatomy of creativity*. New York: Basic Books.

Goldstein, R. (2005). *Incompleteness: The proof and paradoxes of Kurt Gödel*. New York: W. W. Norton.

Hadamard, J. (1945). *The psychology of invention in the mathematical field*. Princeton, NJ: Princeton University Press.

Hall, D. L., & Ames, R. T. (2001). *Thinking from the Han: Self, truth, and transcendence in Chinese and Western culture*. Albany: State University of New York Press.

Hartshorne, C. (1970). *Creative Synthesis and Philosophic Method*. LaSalle, IL: Open Court.

Hausman, C. R. (1975). *A discourse on novelty and creation*. The Hague: Martinus Nijhoff.

Heelan, P. (1970). Quantum and classical logic: Their respective roles. *Synthese, 21*, 2-33.

Heelan, P. (1971). Logic of framework transpositions. *International Philosophical Quarterly, 11*, 314-34.

Heelan, P.(1983). *Space-perception and the philosophy of science*. Berkeley: University of California Press.

Herbst, D. P. (1993). What happens when we make a distinction: An elementary introduction to co-genetic logic. *Cybernetics & Human Knowing, 2* (1), 29-38.

Holton, G. (1978). *The scientific imagination: Case studies*. Cambridge, MA: Harvard University Press.

Hookway, C. (1985). *Peirce*. London: Routledge and Kegan Paul.

Huntington, C. W. Jr. (1989). *The emptiness of emptiness*. Honolulu: University of Hawaii Press.

Iida, S. (1980). *Reason and emptiness: A study in logic and mysticism*. Tokyo: Hokuseido Press.

Isaacson, W. (2007). *Einstein: His life and universe*. New York: Simon and Schuster.

Kalupahana, D. (1986). *Nāgārjuna: The philosophy of the middle way*. Albany: State University of New York Press.

Kauffman, L. H. (2001). The mathematics of Charles Sanders Peirce. *Cybernetics & Human Knowing, 8* (1-2), 133-40.

Kauffman, L. H. (2002a). Laws of form and form dynamics. *Cybernetics & Human Knowing, 9*, (2), 49-63.

Kauffman, L. H. (2002b). Time, Imaginary value, paradox, sign and space. *Computing Anticipatory Systems* (Fifth International Conference. AIP Conference Proceedings 627, pp.146-59). Retrieved November 17, 2009 from: http://www.math.uic.edu/~kauffman/TimeParadox.

Kauffman, L. H. & Varela, F. J. (1980). Form dynamics. *Journal of Social and Biological Structures, 3*, 171-216.

Kaufman, J. C., & Baer, J. (2006). *Creativity and reason in cognitive development*. New York: Cambridge University Press.

Lao Tzu (1963). *Tao Te Ching* (D. C. Lau, trans.). Middlesex: Penguin.

Laycock, S. W. (1994). *Mind as mirror and the mirroring of mind: Buddhist reflections on Western phenomenology.* Albany, NY: State University of New York Press.

Loy, D. (1989). *Non-duality.* New Haven, CT: Yale University Press.

McCagney, N. (1997). *Nāgārjuna and the philosophy of openness.* Lanham: Rowman and Littlefield.

Mandelbrot, B. B. (2004). *Fractals and chaos: The Mandelbrot set and beyond.* New York: Springer.

Mansfield, V. (1989). Madhyamika Buddhism and quantum mechanics: Beginning a dialogue. *International Philosophical Quarterly, 29* (4), 371-91.

merrell, f. (1998). *Tasking textuality.* Berlin: Peter Lang.

merrell, f. (2007). *Processing cultural meaning.* Ottawa: Legas.

merrell, f. (in press). *Entangling forms: Within everyday processes and practices.* Mouton de Gruyter.

Michalko, M. (2001). *Cracking creativity: The search for creative genius.* New York: Ten Speed Press.

Miller, A. (2002). *Einstein, Picasso: Space, time, and beauty that causes havoc.* New York: Basic Books.

Miller, D. L. (1989). *Philosophy of creativity.* New York: Peter Lang.

Nachmanovitch, S. (1990). *Free play: Improvisation in life and art.* New York: Penguin Putnam.

Nāgārjuna (1967). Fundamentals of the middle way. In F. J. Streng (Ed.), *Emptiness: A study of religious meaning* (pp. 1-25). New York: Abingdon Press.

Nicolis, G. & Prigogine, I. (1989). *Exploring complexity: An introduction.* San Francisco: W. H. Freeman.

Odin, S. (1996). *The social self in Zen and American pragmatism.* Albany: State University of New York Press.

Pagels, H. (1988). *The dreams of reason: The computer and the rise of the sciences of complexity.* New York: Simon and Schuster.

Peirce, C. S. (1931-35). *Collected Papers of Charles Sanders Peirce,* (vols. 1-6, C. Hartshorne & P. Weiss, Eds.), Cambridge, MA: Harvard University Press (reference to the collected papers will be designated CP).

Peirce, C. S. (1958). *Collected Papers of Charles Sanders Peirce* (vols. 7-8, A. W. Burks Ed.). Cambridge, MA: Harvard University Press (reference to the collected papers will be designated CP).

Plaw, A. (Ed.) (2005). *Frontiers of diversity: Exploration in contemporary pluralism.* New York: Rodopi.

Pope, R. (2005). *Creativity: Theory, history, practice.* New York: Routledge.

Priest, G. (1987). *In contradiction: A study of the transconsistent.* The Hague: Martinus Nijhoff.

Priest, G. (2004). What's So Bad About Contradictions? In G. Priest, J. C. Beall, & B. Armour-Garb (Eds.), *The law of non-contradiction* (pp. 23-38). Oxford: Clarendon.

Priest, G., Routley, R., & Norman, J. (Eds.) (1989). *Paraconsistent logic: Essays on the inconsistent.* Munich: Philosophia.

Prigogine, I. (1980). *From being to becoming: Time and complexity in the physical sciences.* New York: W. H. Freeman.

Prigogine, I. (1997). *The end of certainty.* New York: Free Press.

Prigogine, I., & Stengers, I. (1983). *Order out of chaos: Man's new dialogue with nature.* New York: Bantam.

Putnam, H. (1971). How to Think quantum logically. In P. Suppes (Ed.), *Logic and probability in quantum mechanics* (pp. 47-53). Dordrecht: D. Reidel.

Putnam, H. (1983a). Vagueness and alternative logic. *Erkenntinis, 19,* 297-314.

Queiroz, J., & merrell, f. (2005). Abduction: Between subjectivity and objectivity [special issue]. *Semiotica, 153,* 1/4.

Rescher, N. (1993). *Pluralism: Against the demand for consensus.* Oxford: Clarendon.

Rescher, N., & Brandom, R. (1979). *The logic of inconsistency: A Study of the non-standard possible-world semantics and ontology.* Totowa, NJ: Rowman and Littlefield.

Robertson, R. (1999). Some-thing from no-thing: G. Spencer-Brown's laws of form. *Cybernetics & Human Knowing, 6* (4), 43-57.

Root-Bernstein, R. & Root-Bernstein, M. (1999). *Sparks of genius: The thinking tools of the world's most creative people.* Boston: Houghton Mifflin.

Rosenthal, S. B. (1994). *Charles Peirce's pragmatic pluralism.* Albany: State University of New York Press.

Rosenthal, S. B. (2000). *Time, continuity, and indeterminacy: A pragmatic engagement with contemporary perspectives.* Albany, NY: State University of New York Press.

Rothenberg, A.(1990). *Creativity and madness.* Baltimore, MD: The Johns Hopkins University Press.

Rothenberg, A., & Hausman, C. (Eds.) (1976). *The creative question.* Durham, NC: Duke University Press.

Sawyer, K. R. (2006). *Explaining creativity: The science of human innovation.* New York: Oxford University Press.

Schiltz, M., & Verschraegen, G. (2002). Spencer-Brown, Luhmann and tautology. *Cybernetics & Human Knowing, 9* (3/4), 55-78.

Sebeok, T. A. (1981). *The play of musement.* Bloomington, IN: Indiana University Press.

Seife, C. (2000). *Zero: The biography of a dangerous idea.* New York: Penguin.

Shlain, L. (1993). *Art and physics: Parallel visions in space, time, and light.* New York: William Morrow.

Stearns, I. S. (1952). Firstness, Secondness, and Thirdness. In P. P. Wiener & F. H. Young (Eds.), *Studies in the Philosophy of Charles Sanders Peirce* (pp. 195-208). Cambridge: Harvard University Press.

Turrisi, P. (1990). Peirce's logic of discovery: Abduction and universal categories. *Transactions of the Charles S. Peirce Society, 26* (4), 465-97.

Weisberg, R. W. (2006). *Creativity: Understanding innovation in problem solving, science, invention, and the arts.* Hoboken, NJ: John Wiley.

Wertheimer, M. (1959). *Productive thinking* (rev. ed.). New York: Harper.

West, T. G. (1997). *In the mind's eye: Visual thinkers, gifted people with dyslexia and other learning difficulties, computer images and the ironies of creativity.* New York: Prometheus.

West, T. G. (2004). *Thinking like Einstein: Returning to Our Visual Roots with the Emerging Revolution in Computer Information Visualization.* New York: Prometheus.

Wheeler, J. A. (1980a). Beyond the black hole. In H. Wolff (Ed.), *Some strangeness in the proportion* (pp. 341-80). Reading, MA: Addison Wesley.

Wheeler, J. A.(1980b). Law without law. In P. Medawar & J. H. Shelley (Eds.), *Structure in science and art* (pp. 132-68). Amsterdam: Excerpta Medica.

Wheeler, J. A. (1984). Bits, Quanta, Meaning. In *Theoretical physics meeting*, (pp. 121-34). Naples: Edizioni Scientifiche Italiane.

Wirth, U. (1999). Abductive reasoning in Peirce's and Davidson's account of interpretation. *Transactions of the Charles S. Peirce Society, 35*, 115-27.

Craft, D. (2009). *For Lou DiMattia*. Digital Collage; 34 x 34 cm.

Cybernetics and Human Knowing. Vol. 16, nos. 3-4, pp. 107-148

The View from Husserl's Lectern
Considerations on the Role of Phenomenology in Cognitive Semiotics

Göran Sonesson[1]

The aim of this paper is to consider in what way semiotics, cognitive science, and phenomenology, which stem from different traditions that have only rarely been known to intermingle, can enter harmoniously into a common research paradigm, phenomenological cognitive semiotics, to which each one of them has a specific contribution to make, In the first part of the paper, the relationship between (traditional) cognitive science and semiotics is elucidated from different points of view. The second part is concerned with methodology in general, and with the phenomenological method in particular, contrasting the conceptions of the latter propounded by Husserl and Peirce. I then go on to discuss some onto-epistemological consequences of opting for the phenomenological method. Finally, I argue that phenomenological cognitive semiotics might play a central part in the renewal of the Enlightenment project.

On Christmas Eve, Edmund Husserl presented his young friend [Jan Patočka] with a special gift. It was a desk-top lectern, crafted of light wood, which Husserl's own friend and mentor, Tomáš Masaryk, had in turn given him a lifetime earlier, in Leipzig in 1878. "And so," as Patočka would write yet a lifetime later, "I became the heir of a 'tradition.' T[he] Enlightenment tradition ..." (Kohák, 1989)

1. Introduction

Given the bad press endured by Husserlean phenomenology during the last century in the Anglo-American cultural sphere, as well as in most of Northern Europe, it may seem paradoxical to associate phenomenology with the Enlightenment tradition. And yet Husserl (1954) took a clear stance in defending the Western cultural tradition in *Krisis*. In fact, what program could be more rationalist than the task Husserl set himself, which consists in rendering explicit to understanding even that which is customarily taken for granted.

To begin with, however, we have more strange bedfellows to contend with. Already the title of this article risks exposing me to the shameful designation of being an eclectic. On the contrary, I intend to show that semiotics, cognitive science, and phenomenology can not only sleep in the same bed, but can also work together while being wide-awake.

The term *cognitive semiotics* has been proposed a number of times – by Daddesio (1996), by Margariños de Morentin on his webpage for about two decades, and by a number of others, including Per Aage Brandt and Jean-Marie Klinkenberg, even

---

1. Department of Semiotics & Centre for Cognitive Semiotics, Lund University.
E-mail: Goran.Sonesson@semiotik.lu.se

before that, at least in oral form. It is not clear that the term has always been used to cover the same thing: Here, at least, we will invoke it in order to endow semiotics with a more empirical mode of operation, and to integrate the fundamental issues of meaning into the framework of the cognitive sciences, which seems particularly relevant nowadays when the latter have become preoccupied with the evolution and development of human beings in relation to other animals (Donald, 1991, 2001; Deacon, 1997). At the same time, there is now a branch of cognitive science that puts its emphasis on consciousness and which, more often than not, derives its inspiration from phenomenology (Thompson, 2007; Gallagher, 2005; Zlatev, 2008). It therefore seems that a phenomenologically inspired cognitive semiotics could be justified already by the common operation of transforming two separate intersections into the union of all three sets.

However, there are more solid reasons for proposing such a synthesis. In fact, many of these reasons have been given in my earlier work. I have been involved with phenomenological cognitive semiotics from the very start of my career without knowing it – or rather, without using the term. My basic position is still the same as the one I defended in my doctoral dissertation (Sonesson, 1978) and indeed in the programmatic article entitled "A Plea for Integral Linguistics" (Sonesson, 1979): that theoretical linguistics requires a general semiotic framework, and that an integrative semiotic theory can only be built on the basis of the phenomenological method in combination with empirical research. In this respect, I was, and still am, less a follower of Husserl than of his direct disciple Aron Gurwitsch, who also strived to take empirical research and existing psychological theories into account, but always confronting them with results of phenomenological reflection. In fact, however, even Husserl himself incorporated the findings and theories of Gestalt psychology into his discussion.[2]

The phenomenological approach in the Husserlean sense amounts to the attempt to render explicit the structuring present in the field of consciousness. After Husserl, of course, there is hardly any way of doing pure phenomenology, in the sense of remaining alone with the phenomena themselves, to the exclusion of all other kinds of company, because there are least also Husserl's writings to contend with. Furthermore, we will have to enter into a dialogue which is just as prolix with writings of quite different inspiration – as I did in my earlier work (Sonesson, 1989), and as Gurwitsch (1957) may have been the first to do, although Paul Ricœur, at least from his metaphor book onwards (Ricœur, 1975), must be considered the emblematic trailblazer of that approach.[3] It is true, as Thompson (2007, p. 267ff) points out, that since Husserl's time, there has been too much writing about Husserl, to the detriment of new applications of the phenomenological method. While I think it is necessary to

---

2. How this is theoretically feasible is of course an issue, but Husserl never discusses this, and even Gurwitsch's attempt to account for it in terms of the difference between phenomenological philosophy and phenomenological psychology is hardly entirely satisfactory.

3. For an introduction to phenomenology from the point of view of contemporary cognitive science, cf. Gallagher & Zahavi, 2008.

return to the things themselves, as the phenomenological slogan goes—and I have been doing some original phenomenological work myself, in particular involving the sign and the *Bildbewusstsein*, from Sonesson, 1989 onwards—I also believe the community of researchers (invoked by Husserl just as by Peirce) has to involved in the process. This is to say that you are not really exclusively confronted with the phenomena, as Husserl at least claimed to be, but you are always engaged in a discussion with other thinkers – and with the world of our experience, as reflected in empirical studies, in the restricted sense usually given to this term, corresponding to third-person-perspective research (Zlatev, 2009). After all, this is what is meant by the notion of tradition.

Phenomenological reflection may thus also be enriched by the reading of other thinkers, who have thought about the same or similar phenomena. In my 1978 book, I was already very critical of French structuralism, but I was also rather sceptical about Peirce. In *Pictorial Concepts* (Sonesson, 1989), I derived much more inspiration from Peirce, and I have continued to do so more and more. But this has happened because I have become convinced that some parts of Peirce's work is actually very good phenomenology. Nevertheless, I still think it is quite arbitrary to decide that everything in the world of our experience pertains to three categories without clearly determining from what point of view, and without admitting that there may be more or fewer categories from other, equally valid vantage points. The proof of Peirce, like the proverbial pudding, is in the eating – that is, in phenomenological analysis. Actually, it is found in what Husserl termed *free variation in the imagination*. Once you state clearly what the principle underlying the three categories is, you can investigate which are its possible variations, and then you can find out whether there are three possibilities or more—or less.

Because of the common misapprehension of phenomenology in Anglo-American and North-European culture, it is often supposed that by basing oneself on phenomenology, you take your departure in common sense. This may be true of ethnomethodology, which claims to take its origin in phenomenology (Garfinkel, 1967), but, in relation to Husserlean phenomenology, there is a fundamental misunderstanding in this interpretation. In one sense, phenomenology is quite the opposite of common sense. It directs our thinking to the kind of thinking done by common sense in order to go beyond it. However, phenomenology may not always be able to do this alone. That is why we need empirical research to discover new things. And we need the critical discussion of other theories. Contrary to what Husserl supposes, you may need such a discussion already in order to discover the kind of parameters to submit to variation in the imagination. This is something that I have recently been doing with respect to biosemiotics (Sonesson, 2006b, 2009; cf. Zlatev, 2009). In the process I have learnt new things from biosemiotics, but I have also discovered in which aspects biosemiotics (as much of Peircean theory) is at fault: in only seeing gray cats even in open daylight. In other words: in treating all kinds of meaning (and perhaps some things which are not even meanings because there is not subject to experience them) as signs, without attending to the differences.

In this paper, I will argue that the resolution of these problems is to be found in the combination of ideas from semiotics, cognitive science and phenomenology. I am very unsympathetic to the idea of taking over wholesale an entire framework, be it that of Peirce, Husserl, or any other thinker. Rather than a scientific process, this would seem to resemble religious conversion. Although any scientific theory must of course be coherent, it has to be acquired piecemeal, each part being separately discussed and experienced from the first-person, second-person, and third-person perspectives (Zlatev, 2009). It should also be made clear that what I have adopted is a special domain of study corresponding to the combination of those domains presently occupied by cognitive science and semiotics, as well as the method known as phenomenology, not any specific theories already known in any of these domains, or derived from this method. In the following, I intend first to explore the domain of cognitive semiotics, and then consider the phenomenological method as such and in relation to the domain referred to, which will also allow me to say something about the onto-epistemological presuppositions of these choices. Finally, I will return to the relevance of the Enlightenment tradition today.

2. The Construction of Cognitive Semiotics

Cognitive semiotics, as it is here understood, aims to bring together the knowledge base and models of cognitive science and semiotics.[4] It seems to have been invented several times over, probably because it is needed. It is needed, first of all, because it is hardly possible to talk about the human mind without also involving the specific ways human beings have of communicating with each other, and vice versa. This has to do with the peculiar intersubjective nature of the human Lifeworld, opposed in that respect to other animal *Umwelten* (cf. section 2.4). What is fundamentally missing in cognitive science is a conception of meaning. In cognitive science, phenomena endowed with meaning are either discussed without being treated as such, or they are not adequately described (see section 2.2). Within the branch of cognitive science concerned with consciousness (e.g., Thompson, 2007) meaning is certainly epitomized as a product of subjectivity, but it is not elucidated in its own right. On the other hand, what seems to be lacking in semiotics, is actual empirical research, in the sense of experiments and observations of relevant situations (cf. section 2.3). Furthermore, phenomenology is necessary, in order to conceive an adequate semiotic theory, as well as in the task of bringing together cognitive science and semiotics. Without the elucidation offered by the phenomenological method, semiotics and cognitive science risk indeed to end up forming an eclectic patchwork.

2.1 Semiotics on Its Own
It still seems to be impossible to establish a consensus among all semioticians on what semiotics is all about; and many semioticians will not even care to define their

---

4. A better term may be *semiotic cognitive science*, which Deacon once used in a lecture, the date of which I cannot now pinpoint. But *cognitive semiotics* now seems to stand a better chance of being established as a notion.

discipline. It is understandable, then, that a semiotician such as Paul Bouissac (1999) may describe semiotics as being chiefly a kind of meta-analysis, and thus consisting "in reading through a large number of specialised scientific publications, selected among the published literature in one or several domains of inquiry, and of relating the partial results within a more encompassing model than the ones that are held by the various specialists concerned"(Bouissac, p. 4). This certainly accounts for a lot of what semioticians do, but it could also be said the cognitive sciences and many other interdisciplinary perspectives. The reading in question, and the ensuing acts of comparison and integration, must be made from some particular point of view. Indeed, it can even be argued that cognitive science, by definition, has been better at doing meta-analysis than semiotics, because it is characterized by the confluence of various earlier research traditions, whereas semiotics has too long been hampered by the autonomy postulate, taken over from Saussurean and Chomskyan linguistics, according to which semiotics, like Saussurean and Chomskyan linguistics, should not be tainted by any commerce with other disciplines.

However, if we attend less to definitions than to real research practice, and if we leave out those would-be semioticians who simply do not seem to be doing anything very new (those who merely go on doing art history, literary history, philosophy, and so on), it seems possible to isolate the smallest common denominators of the discipline (cf. Sonesson, 1996). The subject matter of semiotics, like those of psychology and sociology, does not exist separately somewhere "out there": just like society and the mind, meaning is entangled into everything else, and is abstracted out from it, applying some particular standpoint, or, in other terms, a principle of relevance. The particular point of view of semiotics is to study the point of view itself, as Saussure (1973, see p. 23) once put it,[5] or, in the words of the later Peirce, it is mediation, the fact of other things being presented to us in an indirect way (cf. Parmentier, 1985). In this sense, there is more to semiotics than signs. As is well-known, both Umberto Eco and the followers of Greimas would like to rid semiotics of the notion of sign altogether, whereas Peirceans continue to see signs everywhere. But one can argue that this is so because these schools have different definitions of the sign, which are either vague or not very explicit.

If semiotics is a science, we have to start out by explicating the notion of science as such. As a first approximation, one may want to say that a science is a particularly orderly and systematic fashion of describing and analyzing or, more generally, interpreting a certain part of reality, using different methods and models. At this point we may want to introduce a division between the natural sciences, on the one hand, and the social and human sciences, on the other, which, following a traditional hermeneutical conception echoed by Eco (1985/1988, see p. 351), separates the interpretation of facts from the interpretation of interpretations. Normally, it is added

---

5. This is, as Prieto (1975a) has convincingly shown, the real meaning of the famous Saussurean saying that the point of view creates the object of linguistics, which, as observed by Sonesson (1989, p. 28), is extended to all other "sciences sémiologiques" by Saussure (1974, p. 47). Without referring to Vico (2004), Prieto here seems to take for granted Vico's postulate, which is something that will be discussed later on in the text.

that the first kind of knowledge involves phenomena for which laws may be formulated, while the second kind only refers to unique occurrences; and that while, the first kind can be explained, the second kind may be understood But there is something seriously wrong with this analysis, even at its earliest stage. Not all sciences appear to have their own reserved piece of reality to study. Rather, sciences may be defined either as being concerned with a particular domain of reality, or as applying a particular point of view to the whole of reality (cf. Sonesson, 2004).[6] Thus, French studies are involved with French language and literature, linguistics with all languages (or what is common to all languages); similarly, the history of religions describes a very particular domain of reality, religion, as it evolves through history (and pre-history). Even within the natural sciences, there are some sciences that have their particular domains, such as geography, astronomy, and meteorology. This seems to be even more obviously true of such applied sciences as medicine and dentistry.

But there is no semiotic domain, just as there is no psychological or sociological one. Rather, everything may be studied from the point of view of its semiotic, psychological, or sociological properties. We find the same thing in the natural sciences: Chemistry and physics often appear to be different points of view taken on the very same subject matter. This is not the whole truth: in fact semiotics, psychology and sociology only apply their points of view to the human world, or at least to the world of living beings (in most cases, to animals, not to plants). So the point-of-view approach needs to be supplemented by a domain-approach. The domain of chemistry and physics is much wider: its goes well beyond the human world. But both apply the same point of view to the human world and what lies behind it, which is impossible for semiotics, as well as for psychology and sociology. Contrary to chemistry and physics, biology is not just another point of view, but it is also domain-specific: It only involves living creatures. This may explain that there is now such a speciality as biosemiotics but not (or at least I hope not) chemical semiotics.

Semiotics, it was suggested above, is a science, the point of view of which may be applied to any phenomenon produced by human beings as well as by other animals. As such, it is concerned with the different forms and conformations given to the means through which animate beings take themselves to have access to "the world." In studying these phenomena, semiotics should occupy the *standpoint* of human beings, or, when this is relevant, of some particular part of humankind – or, again, to the extent that this is possible, of the species involved. Indeed, as Saussure (1974, p. 47) argues, semiotic objects exist merely as those points of view that are adopted on other, material objects, which is why these points of view cannot be altered, according to Saussure, without the result being the disappearance of the semiotic objects as such. Saussure voices this claim, because he wants to make the object of general linguistics (and the other semiological sciences) appear out of the background of the different sciences existing at his time which were involved with language. That the object of

---

6. This seems to fit in very well with Peirce's conception, according to which "the only natural lines of demarcation between nearly related sciences are the divisions between the social groups of devotees of these sciences" (CP 8.342), themselves determined by the kind of problems they are intent on resolving

semiotics may disappear again is testified by the practice of most of contemporary cognitive science. However, Saussure's claim should not be taken as a nominalist profession of faith. As shown by Prieto (see 1975a, p. 144; 1975b, p. 225f), Saussure's phrase can, in the given context, only be interpreted as concerning the reconstruction of the point of view of the speakers. The task of semiotics is thus not to develop any particular philosophical position, but to construct a model of the users of these meanings in their on-going practice of the Lifeworld. We cannot, like the philosopher Nelson Goodman (1968), reject the folk notion of picture because of its incoherence, but must discover its peculiar systematicity. From a semiotic point of view, it does not matter whether the researcher favours a *nominalist* view of reality, or some other conception. It is the Lifeworld notion of life that we must reconstruct, and the Lifeworld notion is certainly not (purely) nominalist. Even a nominalist must somehow accept that concepts and ideas exist, in order to live and act as a member of human society, and a semiotic description of the thinking of a nominalist could not be phrased in nominalist terminology.

Nor do I think it makes much sense to claim, with Umberto Eco (1985/1988, p. 323ff) that, on the one hand, there are certain specific semiotic sciences, such as those which study the interpretative habits of events in verbal language, gestures, traffic signs, pictures, and so on; and, on the other hand, there is a general semiotics, which simply postulates the concept of sign, thus permitting us to speak about superficially dissimilar things within a unified framework. Curiously, Eco even claims that the fact of there being different semiotic points of view demonstrates that semiotics is a philosophical activity; but, at the very least, this would show that semiotics is a variety of different philosophical and/or scientific activities. Actually, a more adequate conclusion would be that semiotics—just as sociology, psychology, archaeology, literary history, and so on—can be practised from the point of view of different philosophical conceptions. Thus, there may be a structuralist semiotics, a nominalist semiotics, a phenomenological semiotics, and so on—just as there may be, for instance, a processural and a post-processural archaeology, a positivist and a postmodern art history, and so on.

Generalizing from the case of linguistics, Prieto (1975a) takes it to follow that semiotics must be restricted to the knowledge shared by all users of the system. Pursuing the same analogy, we are bound to realize that it is necessary to descend at least one level of analysis below the level of which the user is usually aware in order to take account of the presuppositions underlying the use of the system. This is necessary in order to explain the workings of such operative, albeit tacit, knowledge that underlies the behavior constitutive of any system of signification. It goes without saying that this knowledge must, in principle, be accessible to consciousness, without which the phenomenologist, who is thrown upon his own ordinary human resources, would be unable to reach it (cf. Sonesson, 1989, p. 26ff; 2007).

Semiotics attends to all phenomena considered in their qualitative aspects rather than the quantitative ones, but it nevertheless aims at formulating rules and regularities, rather than being reduced to the interpretation of unique objects. This is to say that all semiotic sciences, including linguistics, are *nomothetic* sciences, concerned with generalities, not idiographic sciences, such as art history and most of the humanities, which take as their object an array of singular phenomena, the common nature and connectedness of which they take for granted. Just like linguistics, as understood in the Saussurean tradition, but contrary to the natural sciences and the social sciences (according to most conceptions), semiotics is concerned with qualities, rather than quantities – that is, it is concerned with categories more than numbers. Thus, semiotics shares with the social and natural sciences the character of being a law-seeking, or nomothetic, rather than an idiographic, science, while retaining the emphasis on categories, to the detriment of amounts, which is peculiar to the human sciences. Semiotics is not restricted to any single method, but uses a plurality of such, varying from the analysis of concrete texts (text analysis), to classical experimental technique, and imaginary variation reminiscent of the one found in philosophy (system analysis). Moreover, the construction of models is a peculiar feature of the semiotic standpoint, if it is compared to most of the humanities (which is not to say that these models must be taken over from linguistics, as is often believed). Indeed, semiotics differs from traditional approaches to *humanitas* in employing a model that guides its practitioners in their effort to bring about adequate analyses, instead of simply relying on the power of the *innocent eye*. Two very general categories of models could be taken to be the analytical and the synthetic ones, but it might be more to the point to observe that most real models have analytical and synthetic aspects. Science normally makes its analysis by means of synthesis, that is, a tentative synthesis, which may then have to be modified in the confrontation with the object, analysed.

In actual practice, however, there have been preciously few experiments in semiotics. According to many exponents of contemporary semiotics (e.g., Greimas, 1970), semiotics is a pure, or autonomous, science, similar to structural linguistics (Itkonen, 1978). Other researchers, notably in the United States, tend to look upon semiotics as a meeting-place of many different sciences, a kind of interdisciplinary framework common to the humanities and the social sciences, including, by some accounts, biology and neurology. My approach is different from both those characterized above: I take the results of all disciplines involved with the same subject matter (i.e. meanings, signs, words, gestures, pictures, photographs, etc., as the case may be) to be relevant to semiotics, but only once they have been reviewed, redefined and complemented from a specifically semiotic viewpoint. Unlike most of the venerable semiotic tradition, I have always argued against the autonomy postulate, basing my own work to a large measure on an interpretation of experimental results (most notably in Sonesson, 1989). It is only in recent years, however, that have I been involved with the design of experiments myself. Scholars such as René Lindekens and Martin Krampen, who already in the heyday of structuralism set up their own experimental studies, basing themselves on semiotic models, may be seen as precursors of experimental, and therefore at least in one sense, cognitive, semiotics. Using such methods, Lindekens (1971, see p. 178ff) showed that the interpretation of a photograph is changed according as it has been made more or less contrasted or nuanced in the process of development. Espe (1983a, 1983b), in a similar set-up, demonstrated interesting interactions between different factors, with the general result that an identical photograph may carry very different affective import when being differently contrasted. Krampen (1991) used experimental methods in the study of children's drawings. In domains of semiotics which have developed into specialities of their own outside of semiotics, such as gesture studies, there has of course been much more experimental work, as there obviously has been in (psycho)linguistics. In general, however, this is an aspect in which semiotics has a lot to learn from the cognitive sciences, as well as from some of the sciences anterior to the cognitive synthesis, such as cognitive and perceptual psychology.

2.2 The Three Ages of Cognitive Science

It no longer makes sense to invoke cognitive science as a whole. Cognitive science can be practiced, and indeed has historically been practiced, from very different points of view. There is some paradox to the very name *cognitive science*, because its initial aim was to do away with cognition, and indeed consciousness, as we know it. Indeed, the fact that mental life could be simulated on a computer was supposed to show that mental notions could be dispensed with altogether. Consciousness was taken to be no more than a set of calculations based on some snippets of code made by the human brain. Jerry Fodor's (1987) argument for the language of thought is the most explicit version of this point of view. This conception is still very influential within cognitive science in the form of Daniel Dennett's (1987) idea about the intentional stance: that human beings simply work like computers, with the added twist that they, for no useful reason at all, happen to think they are conscious.

In the second age of cognitive science, some researchers realized that human beings (as well as, on some interpretations, some computer programs) could not function outside of a human lifeworld and without taking their bearings from their outside bodily form. This brings us to the notion of situatedness, which has henceforth played an important role in cognitive science, and to the complementary notion of embodiment. These notions served to bring ideas from phenomenology and other traditions involved with consciousness into the fold of cognitive science. Before this moment many phenomenologists and philosophers of consciousness—most famously Hubert Dreyfus (1992) and John Searle (1997, 1999)—were violently opposed to cognitive science, a fact that hindered any cross-fertilization. However, both situatedness and embodiment can be given—and have been given—other, more mechanistic, interpretations. The preoccupation with notions such as agency, intentions, consciousness, empathy, intersubjectivity, and so forth, remain atypical of cognitive science as a whole, though they are a major topics within consciousness studies, such as practised by Evan Thompson (2007), Shaun Gallagher (2005), Dan Zahavi (Gallagher & Zahavi, 2008), and a few others.[7] In fact, these notions are anathema to much of cognitive science, both in its classical version and, in a more implicit and confused way, in what must still be described as mainstream cognitive science, associated with the work of Lakoff and Johnson (1999), Dennett, Fodor, etc.[8]

To Lucy Suchman (1987) and her followers, the term *situated* expressed a need to take context into account. This applies to embodiment as well, because our own body is the primary context of all our actions. *Embodiment* is a more precise term than *context* and perhaps *situatedness* can be defined more precisely too. In any case, even if *situated* and *embodied cognition* are fashionable terms at present, mainstream cognitive science still does not seem to take them in the direction of consciousness studies. However, in the second tradition of cognitive science, the body which forms this context is not the body as lived, that is, as a meaning, but the physical body as studied in neuroscience. Lakoff, Johnson, and their followers today form the core of what is meant by mainstream cognitive science. Although their work is extremely confused and contradictory, which is testified most clearly by their different levels of embodiment, which are distinguished and then conflated (as shown most clearly by Zlatev, 2005, 2007), and by the incoherent attacks on Western philosophy (cf. Haser, 2005), and even though it contains superficial references to part of the phenomenological tradition, a close reading of, in particular their most recent publications, shows that in actual fact they are back at a conception identical in practice to that of classical cognitive science, with the brain substituted for the computer. The body they are talking about is reduced to the neurons and synapses of the brain. Thus, embodiment, in this tradition is certainly not part of context. This is equally true if their work is interpreted in terms of the kind of influence they have had.

Another related problem derives from the term *cognitive* as such, as it appears in the name of the enterprise. In the traditional discipline of cognitive psychology, and in the psychology of development, as, for instance, in the Piaget tradition, the term *cognitive* has a rather clear, well circumscribed meaning, being opposed, notably, to perception, unconscious processes, and probably empathy in most senses of the term. At least prototypically, or as a goal state, it involves rational operations, such as those that are characteristic of argumentation or problem solving. Although I am not aware of any explicit definition of the term within cognitive science, it is clear that the term *cognitive* here has taken on a much vaster, fuzzier meaning: originally, it corresponded to everything which could be simulated by a cognitive device such as a computer, and nowadays, it appears to stand for anything which can be localized in the brain. According to the language-of-thought hypothesis (first formulated by Fodor), even categorical perception and other elementary perceptual operations are based on cognition. Contemporary representatives of cognitive sciences such as Lakoff and Johnson would seem to claim that also thinking, in a more traditional sense, might be

---

7. In Sonesson (2007), I took Varela, Thompson, & Rosch (1991) to task for having seriously misunderstood Husserl and attributing many of his real accomplishments to Merleau-Ponty; so it is interesting to note that in his new book, Thompson (2007) does not only assign a much more preponderant part to Husserl, but includes an appendix of *mea culpa* specifying the respects in which he and Varela were wrong about Husserl, having being basically mislead by the interpretation proposed by Dreyfus.

8. My first tradition seems to correspond to what Thompson (2007, p. 4ff) calls cognitivism, but the other two only overlap somewhat with Thompson's connectionism and embodied dynamicism.

reduced to very simple operations, in which case *cognitive science* becomes a misnomer. If the first tradition of cognitive science thus reduced the mind to a computer, the second tradition introduced a new kind of reductionism involving the brain, so it is only the third tradition that holds out some hope for an approach to meaning.

The kind of cognitive science with which I here would like to organize an encounter is that of the second tradition, that is, the brand whose real epistemological horizon is phenomenology, in its classical Husserlean form as well as in its recent versions within consciousness studies; and with Searle, whose version of the philosophy of mind is to a large extent either crypto-phenomenological or a parallel development arriving at the same general conclusions.

2.3. The Semiotic Turn in Cognitive Science and Vice-versa

Like cognitive science, semiotics is often conceived as an interdisciplinary perspective that has occasionally gained the position of an independent discipline—no doubt less often than cognitive science. Curiously, it might be argued that cognitive science and semiotics cover more or less the same domain of knowledge; or rather, they take a very similar perspective on the world: both are concerned with the way in which the world described by the natural sciences appears to humans beings and perhaps also to animals. Cognitive science (explicitly only in that of the third generation) puts the emphasis on the place of the appearance of this world, the mental domain and on its characteristic operation, consciousness, in its various manifestations; and semiotics insists on the transformations that the physical world suffers by being endowed with meaning.

Semiotics would have little to offer cognitive science if it were only a model or a method, or a philosophical standpoint. But as argued above, semiotics cannot be considered simply in these terms, nor is it simply a critique of ideology or, in Paul Bouissac's (1999) term, a *meta-analysis*. Semiotics must be taken to be a science in its own right (cf. Sonesson, 1992, 1996, 2004). The most obvious reason for this is that semiotics, if it is not erroneously identified with French structuralism, has been using many different models and methods, as well as being practiced from different philosophical points of view. Likewise, it is not simply a meta-analysis or some other kind of interdisciplinary perspective, because that does not tell us anything about its originality. It is interdisciplinary and meta-analytical with a twist: Semiotics takes meaning as its perspective on the world. What this means, no doubt, will remains somewhat obscure until meaning has been phenomenologically elucidated (as in Sonesson, 2006a, in press a, in press b)

The disciplinary history of these two approaches has been very different. Cognitive science is often described as the result of joining together the knowledge base of rather disparate empirical disciplines such as linguistics, cognitive psychology, philosophy, biology, and computer science. Thus, instead of one research tradition connected through the ages, cognitive science represents a very recent intermingling of several research-traditions having developed separately until a few decades ago.

Semiotics has, in a more classical way, developed out of the amorphous mass of philosophy, and still has some problems encountering its empirical basis. It might be suggested that the basic concept of semiotics is the sign, whereas that of cognitive science is *representation*—even though there is a long tradition in semiotics for rejecting the sign concept, as, for instance, in the work of Greimas and his followers (cf. Greimas, 1970; Greimas & Courtès, 1979), and recent cognitive science has distanced itself from the notion of representation (Varela, Thompson, & Rosch, 1991). From the point of view of methods, semiotics is generally speaking stuck between the analysis of single texts and theory construction, whereas cognitive science is closer to relying on experimental methods (including computer simulation). These differences may partly explain why semiotics and cognitive science are rarely on speaking terms.

On the other hand, there have recently been some encouraging developments within cognitive science which, no doubt with some exaggeration, may be qualified as a semiotic turn: there has been a recent interest in meaning as such, in particular as it has developed ontogenetically and phylogenetically, in the human species and, to some extent, in other animals. Terrence Deacon (1997) is a researcher in neuroscience whose work has been particularly acclaimed within cognitive science. Yet he has chosen to express some of his main arguments in a terminology taken over from Peirce, who is perhaps the principal cultural hero of semiotics.[9] Not only Deacon, both other scholars interested in the specificity of human nature now put their emphasis on the concept of sign (which they normally term *symbol*). This is true, in a very general sense, of Donald's (1991) stages of episodic, mimetic, mythic and theoretical culture. It seems to apply even more so to Tomasello (1999), less because of his epigraphs taken from classical semioticians such as Peirce and Mead as well as Bakhtin and Vygotsky, than because of the general thrust of his analysis, which consists in separating true instances of interpreting actions as having a purpose from those which may merely appear to be of that kind. Building on such research Zlatev (2003, 2007, 2008) has explicitly investigated the conditions for the emergence of higher levels of meaning involving mimesis and language, from more basic ones, characteristic of all biological systems (life forms), such as cues and associations.

Interestingly, there has also been an attempt at a true cognitive turn in semiotics, most clearly represented by Daddesio (1995), who tries to absorb the empirical knowledge base of cognitive science into semiotics, siding in this respect with the third generation's way of dealing with consciousness, though he mistakes Lakoff and Johnson for its representatives. His main argument for having recourse to cognitive science, is that the study of signs and sign systems, privileged by semiotics, has to be complemented by investigations of the ways of having access to these signs, which are more properly studied by cognitive science. It is easy to agree with this idea. Unfortunately, however, Daddesio puts the physicalist reductionism of behaviorism

---

9. Without trying in any way to diminish Deacon's contribution—in fact, I find him very convincing whenever he is not having recourse to semiotic terminology—I have earlier expressed serious misgivings about his way of using Peircean terms, because this serves to obscure both the central issues of semiotics, and those introduced by Deacon (cf. Sonesson, 2006b).

characteristic of much of American semiotics on a par with the recognition, on the part of the tradition of Saussure, Cassirer, Husserl, the Prague school, and others, that there is also a third level of meaning, the social, intersubjective, one—which does not exclude the mental world, simply because it is a product of the interaction of many mental worlds. What in the first case is a clear case of denial, is in the second case merely neglect or even a lack of focus.[10] Daddesio would seem to associate semiotics with a particular philosophical standpoint. But this is a point of view that cannot be sustained, as we saw above (2.1). No matter how deservedly different semioticians are faulted with neglecting the relationship between signs and meanings, on one hand, and the ways in which we may access to them, on the other, Daddesio has certainly pointed to an important domain of study which on the whole must be considered to be disregarded, both within semiotics and within cognitive science (as well as by predecessors to the latter within cognitive and genetic psychology): the correlation of signs and other artefacts endowed with meaning, on the one hand, and the modalities of the subject having access to them.

2.4 The Need for Cognitive Semiotics
In order to determine the importance of cognitive semiotics, it is useful to start out from Daddesio's main argument, both in the respect he may be argued to be right, and according to the criticism I have directed as his observation above. If we attend to the behaviorist strand in semiotics, Daddesio is right in emphasizing the importance of studying the relation between signs and meanings and the way in which they become available to us through consciousness. If, on the other hand, we attend to the other tradition in semiotics, which presents signs and meanings as being fundamentally intersubjective structures, then there is still the question involving the ways in which these structures come to be accessible to individual minds, and, in addition, the issue of the emergence of intersubjectivity as such looms large. In both cases, it would seem these problems are exacerbated as soon as the evolution and development of human beings and other animals become relevant to our pursuit.

It could be argued that the founding-fathers of semiotics have already attended to the relationship between signs and the way they are accessible to consciousness. Saussure, for instance, talked about semiotics (his semiology) as being part of social psychology, and he made numerous references to *social intercourse* (in English in the text). But that is about as far as Saussure's interest in these relationships goes. Peirce, of course, included mind in the definition of the sign. But, as is well-known, Peirce gave a very special meaning to the term *mind*, as evidenced by the fact that he thought it (or, at least, the *quasi-mind*, as he was wont to say) did not need to be wedded to any kind of consciousness at all. It is true, no doubt, that Peirce had a lot to say about themes that would usually be considered to be the business of psychology and cognitive science, as is thoroughly documented by Colapietro (1989). And yet, as

---

10. I am of course simplifying the issue: thus, there is a notable ambiguity in the work of Saussure between a social and an outright formalist interpretation.

Peirce neglects the specificity of consciousness, he is unable to elaborate on the relation between signs and the way we as animate beings have access to them. To be more precise, he does not tell us anything about how the relation between consciousness and signs is different from other kinds of relations (notably those between different signs).[11]

As noted above, interactions between subjects and signs become more topical than ever as soon as evolution and development enter the discussion. Contemporary studies of evolution suggest that not only human language, but also the capacity for using pictures, as well as many kinds of mimetic acts and indices, are (at least in their full, spontaneously developed form) uniquely human. It is clear that semiosis itself must be manifold and hierarchically structured, in ways not yet dreamt of in our philosophy. Merlin Donald (1991, 2001) has proposed an evolutionary scale (with analogies in child development explored by Nelson, 1996 & Zlatev, 2003), where the stages of episodic, mimetic, mythic and theoretic culture correspond to types of memory (Cf. Fig. 1.). According to this conception, non-human apes, which otherwise live in the immediate present, are already capable of *episodic* memory, which amounts to the representation of events in terms of their moment and place of occurrence. The first transition, which antedates language and remains intact at its loss (and which Donald identifies with *homo erectus*) brings about *mimetic* memory, which corresponds to such abilities as tool use, miming, imitation, coordinated hunting, a complex social structure and simple rituals. Only the second transition brings about language with its *semantic* memory, that is, a repertory of units, which can be combined. This kind of memory permits the creation of narratives, that is, mythologies, and thus a completely new way of representing reality.

Interestingly, Donald does not think development stops there, even though there are no more biological differences between human beings and other animals to take account of (however, the third transition obviously would not have been possible without the attainment of the three earlier stages). What Donald calls *theoretical culture* supposes the existence of external memory, that is, devices permitting the conservation and communication of knowledge outside human mind (although, in the end, it is of course only accessible to a human mind; cf. Sonesson, 2007). The first apparition of theoretical culture coincides with the invention of drawing. For the first time, knowledge may be stored externally to the organism. The bias having been

---

11. This is certainly not to say that Peirce does not have an idea of how we have access to signs (as an anonymous reviewer reminded me). One of Peirce's most well-known metaphysical claims is that man is a sign (i.e. a symbol). Since human beings are themselves signs they can be attracted to signs, enter into sign communities, use signs, think in signs etc. (cf. CP 7.583). Indeed, Peirce observes in numerous of other ways that human beings and the world are of the same stock, notably when he assures us that we will end up discovering the truth in the long run (i.e. reach the final interpretant, cf. EP 502f). This is an interesting generalization of Giambattista Vico's hermeneutic postulate (cf. Vico, 2004), according to which we can understand other human beings and the works produced by them, that is, the subject matter of the human and social sciences, but not the objects studied by the natural sciences. Characteristically, Peirce denies this difference. But even supposing that we can make sense of the claim that human beings are signs, this does not tell us anything about the specificity of the relation between human beings (and notably consciousness) and signs as compared to the relations between two words and the like.

shifted to visual perception, language is over the millennia transferred further to writing. It is this possibility of conserving information externally to the organism that later gives rise to science.

Figure 1: Donald's Model of Evolution Related to the Notion of Sign Function, as Well as Sign Systems and Embodied Signs.

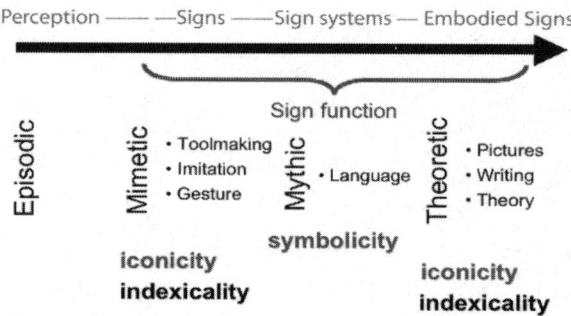

The diverse manifestations of Donald's second stage (mime, skill, imitation and gesture) are, in my view, (at least in part) iconic (based on similarity)—but for the most part they are *tokens* conforming to a *type*—members of a category—not yet signs. Somewhere in between mimesis and language the sign function arises (though Donald notes this only obliquely, mentioning the use of intentional systems of communication and the distinction of the referent). In fact, a lot seems to happen within the stage that Donald's calls mimesis, which is not accounted for, notably because Donald does not pause to consider the partial overlap with Piaget's semiotic function. We have to go from bodily acts as such which are not normally iconic to acts that are imitative and from direct forms of imitation to different kinds of representative imitation.[12] Finally, the fourth stages give rise to organism-independent artifacts, such as, notably, pictures and scale models—as well as writing and theories. Thus, with the advent of pictures, signs reach a post-linguistic stage, which is again iconic. Writing, on the contrary is largely symbolic. Theories, however, may also require an important iconic element, as Peirce has insisted, which is something that Stjernfelt (2007) has recently focused on in his interpretation of Peirce.

---

12. Zlatev (2007, 2009) has suggested a subdivision of the mimetic stage into proto-mimesis, dyadic mimesis, triadic mimesis and (on the borderline to the next stage) post-mimesis. Sonesson (in press b) discusses the relation between imitative acts that are not signs and those which are and the intermediate stage of imitating in order to learn.

Table 1: Donald's Memory Types Analyzed in Relation to the Nature of Accumulation

Type of memory	Type of accumulation	Type of embodiment
Episodic	Attention span (event in time/space)	—
Mimetic	Action sequence co-owned by *Ego* and *Alter*	Own body
Mythic	Transient artefact co-produced by *Ego* and *Alter*	In the interaction between *Ego* and *Alter*
Theoretic	Enduring artefact co-externalised by *Ego* and *Alter*	External in relation to *Ego* and *Alter*

Note: in the sense of Lotman et al., 1975; cf. Sonesson 2007

The stage preceding the attainment of the language capacity requires memory to be located in the subject's own body (see table 1). But, clearly, it can only function as memory to the extent that it is somehow separable from the body as such. The movement of the other must be seen as distinct from the body of the other in its specificity, so that it can be repeated by the self. This supposes a distinction between *token* and *type* (that is, relevance) preceding that of the sign function.

The stage following upon language supposes the sign to acquire a "body" of its own, that is, the ability to persist independently of human beings. Language only seems to require the presence of at least two human beings to exist: they somehow maintain it between themselves. But it is not enough for two persons to know about a picture for it to exist: there must be some kind of organism-independent artefact on which it is inscribed. The picture must be divorced from the bodies (and minds) of those making use of it. Writing is of course, by definition, the transposition of language to independent artifacts. The case of theory may be less obvious: why should not two persons be able to entertain a theory between them? As Husserl (1962) noted well before Donald, complex sign systems, such as mathematics and logic, only seem to function as such when given an existence independent of human organisms. In the case of pictures, Ivins (1953) has observed that it is their reproducibility (as in Floras, for instance) that makes them into scientific instruments. In their capacity of being permanent records, pictures are not, as art historians are wont to say, unavoidably unique, but, on the contrary, are destined for reproduction. Indeed, they permit repeated acts of perception, as do no earlier memory records. This is of course what is known, mainly in Marxist literature, as the process of *reification*. As shown by Cassirer (1942, p. 113ff), this process, far from being only a "tragedy of culture," is a prerequisite for (human) culture.

To understand the specificity of human culture, therefore, we need to explore the interplay of subjects and signs. This is at least one of the reasons why cognitive semiotics is needed.

3. From Methodology to Onto-epistemology

In the constellation of phenomenological cognitive semiotics, phenomenology, as I noted above, enters, not as a substantive domain, but as a method. Not to overburden our argument, let us define a method, in a way which is no doubt too simple, but is sufficient for our purpose, as a series of operations which might be applied in ordered stages to an object of study, with the goal of yielding information of a particular kind about the object studied; and let us similarly decide that a model is a simplified, but still more or less iconic, representation of the object studied which can be more easily manipulated than the real thing, and which (ideally) has the advantage of representing classes of objects of a particular category, rather than a single object, so that, when methodological operations are applied to it, it yields information about the category of objects concerned.

In the previous section, I argued that semiotics cannot be considered to be a method or a model, since, just as all other sciences, it makes use of a wealth of models, as well as a panoply of methods. When one particular model and/or method is attributed to semiotics, it is obviously being confused with one of its manifestations having course during some particular period, most probably the movement known as French structuralism, which was popular in the 1960s and 1970s, but which has since lost its relevance in most quarters, although it survives in the Greimas school, which still has a big following in Latin America as well as in Southern Europe. It may rightly be said about French structuralism that it tried to apply a linguistic model (itself abusively derived from the linguistic structuralism developed, notably, by Saussure and Hjelmslev), as well as to implement (but completely failing to do so) the method of the same linguistic school.

Semiotics, as I suggested above (in section 2), is not restricted to any single method, but is known to have used several kinds, such as text analysis, classical experimental techniques, system analysis, and text classification (cf. Sonesson 1992, 1996).[13] Nor is semiotics necessarily dependant on a model taken over from linguistics, as is often believed, although the *construction of models* remains one of its peculiar features, if it is compared to most of the human sciences. Indeed, semiotics differs from traditional approaches to *humanitas*, whose domain it may partly seem to

---

13. Bouissac (1999a, b) also talks about four ways of acquiring knowledge within semiotics and elsewhere, which partly correspond to my division: experiment and reasoning have obvious parallels, serendipity would for me be something occurring at certain moments within the other strategies, and meta-analysis is a special kind of system analyses, which was already mentioned in part 1.

occupy, in employing models that guide its practitioners in their effort to bring about adequate analyses, instead of simply relying on the power of the innocent eye. After having borrowed its models from linguistics, philosophy, medicine, and mathematics, semiotics is now well on its way to the elaboration of its proper models (cf. Sonesson, 1992, 1993, 1994, 1996, 1998).

2.1 Phenomenology Within a Plurality of Methods

There is, however, another way of looking at different methods, which concerns their various modes of access to knowledge. Zlatev (2009, p. 179) has argued for a form of triangulation between methods that can be categorized on the basis of the type of perspective that the researcher takes to the "data": First-person, Second-person, and Third-person methods (Cf. his table 1 reproduced here as table 2. Also cf. Thompson, 2008, pp. 303ff, 338ff; Gallagher & Zahavi, 2008, p. 13ff).[14]

Table 2: Examples of Methods, Grouped in Terms of Type of Perspective, Used in Developing a Synthetic Cognitive Semiotic Theory

Perspective	Method	Appropriate for the study of
First-person	Conceptual analysis Phenomenological reduction Imaginative (eidetic) variation	Normative meanings, rules Perception Mental imagery
Second-person	Empathy Imaginative projection	Other persons (e.g. as in conversation analysis), "higher" animals
Third-person	Experimentation Brain imaging Computational modelling	Isolated behaviours (e.g. spatiotemporal utterances) Neural processes

Note: Reproduced from Zlatev, 2009

As should be obvious from Zlatev's table, different methods are called for depending on the nature of the phenomena studied. Thus they are used to obtain different kinds of knowledge. First-person methods are required for studying (the contents of) consciousness, second-person methods for studying other subjects, and third-person methods for so-called detached observation. Phenomenology is simply not in competition with experimental studies. As Zlatev rightly observes, "phenomenology was never intended as a method for providing *explanations*, e.g. to answer *why* cats and not rocks appear to us (at least in Western cultures) as conscious beings, and even more so of *causal* explanation, e.g. what neural processes appear to be casually

---

14. As Thompson (2008, p. 301ff) notes, Third person methods should really be called objective methods, because they involve no person at all. This fits in nicely with the Benveniste/Tesnière construal of the third person as a non-person, which I have tried to generalize to cultural semiotics (Sonesson, 2000b).

necessary for consciousness" (Zlatev, 2009, p. 174). However, once you accept (as you most do being a phenomenologist) that there is more than one scientifically viable method, you have to admit that any particular methodological operation must always be accomplished from a specific perspective rather than with what is often desribed as a view from nowhere. This should make you realize the importance of rendering this perspective explicit.

Two caveats, however, seem to me to be necessary. The different methods may be applied to what is, from some point of view (notably a third-person perspective such as that of the natural sciences) described as being the same object or situation, but the kind of knowledge yielded will always be different, and therefore correlation or triangulation will never be straightforward. The most clear-cut illustration of this fact is the use of *protocols* (to use the term introduced by classical cognitive science) written by the experimental subjects during an experimental test session, in which case the knowledge derived from the test is different from that contained in the protocol, although the two kinds of knowledge may profitably be related. Another case in point is the program of neurophenomenology (cf. Thompson, 2008, pp. 329ff, 349ff, etc.), where phenomenological reports are put in relation to the results of brain scanning.

The other point is just as important. To the extent that we are involved with scientific methods, all the knowledge obtained will be of the objective or intersubjective kind (in the sense of not being merely subjective, as the term is used in ordinary language). Although the Lifeworld, and notably ordinary perception, is always subjective-relative, the structures of the Lifeworld, as Husserl insisted, are not themselves relative. If we are concerned with normative meanings /and/ rules, these meanings are of course relative to a particular socio-cultural Lifeworld, but they are objective to anybody who is part of that Lifeworld. However, the kind of introspection that is geared to personal contents of consciousness is not directly at issue in any scientific approach and thus is not accessible to phenomenology. In a parallel fashion, Second person methods are only relevant to science as a way of discovering the objective structures of interaction between different subjects. Terms such as *empathy* and *imaginative projection* would probably more ordinarily be applied to methods which are used by one subject in order to understand another, and as such they are essential, but not scientifically relevant. Of course, such *Lifeworld methods* (or perhaps better, operations) have their structure, and thus become relevant as objects of study to a phenomenological analysis. Indeed, from this point of view, the distinction between first-person and second-person methods appears difficult to maintain, although their opposition to third-person methods is beyond doubt.

Again, a number of glosses must be made on this observation. It could be said that, from a Lifeworld perspective, the modes of existence of the three kinds of knowledge are clearly distinguishable, the first existing for the subject, the other for the subject in his/her interaction with another subject, and the third being independent of any subject. Correlatively, it would seem that there are three *Lifeworld operations*, involving the knowledge of yourself, the knowledge of the other, and finally the knowledge of objective reality, which underlie the scientific methods with which we are concerned, the latter thus only being accessible to those who have beforehand accomplished the corresponding operations of the Lifeworld.[15]

It is important to observe specifically (as was mentioned in passing above when discussing the objectivity of the structures of consciousness) that phenomenological description may be directed at the general structures of consciousness, which are presumably common to all human beings, or at structures of a lower degree of generality, such as those peculiar to a particular socio-cultural Lifeworld, such as the norms governing language use, the habits of gesture, and the like, in one culture as opposed to another.

In fact, some of the methods of analysis employed in linguistics are really phenomenological in nature, though the fact that this is not acknowledged, nor even reaches the awareness of the practitioners, may be detrimental to the results obtained (Sonesson, 2007; Zlatev, 2008). Formally, ideation is very similar to commutation (to use Hjelmslev's term for a procedure well-known to early 20th century phonology, from Prague to Paris), the difference only being that the variation is applied to one plane of language, to see whether this has consequences for the other plane (as noted already by Sonesson, 1989). Also the variations applied to sentences in generative or cognitive linguistics, to decide whether they are still grammatical or not, are of the same general kind. Formally, then, this is the same procedure employed by phenomenology, with the difference that the object to which the procedure is applied has a much lower degree of generality. But ideation does not concern what is necessary in relation to another language plane; basically, it is intrinsic to such a plane. Ideation concerns that which is presupposed by linguistic (and other) analysis. Thus, in relation to linguistic studies, ideation would not ask which content corresponds to which expression, or what sentences are part of the class of all sentences which are grammatical in a particular language, but what properties must be present for something to be a sign, a linguistic sign, a sentence, a discourse, a language, and so on. In this sense, the work of Bühler (1934), Hjelmslev (1959, 1973) and Coseriu (1978) are excellent examples of phenomenology applied to language.

Actually, it may be useful to exercise a little more caution here. The task of phenomenology, as Husserl saw it, was to explain the possibility of human beings having knowledge of the world; as a philosophical endeavour, phenomenology is about the way the world of our experience is constituted. As a contrast, psychology is not about the world, but about the subject experiencing the world. However, every finding in phenomenological philosophy, Husserl claims, has a parallel in phenomenological psychology, which thus could be considered a tradition within psychological science (cf. Husserl, 1962; Gurwitsch, 1974). If consciousness is a relation connecting the subject and the world, then phenomenology is concerned with

---

15. This paragraph would seem to touch, somewhat obliquely, on the old quarrel within hermeneutics whether the operation called *Verstehen*, in Abel's words, is merely a psychological act, or something which intervenes in a particular way in the ascertaining of knowledge (cf. Radnitzky 1970, Vol. II, pp. 26ff, 100ff; Ferris, 1988/2002, p. 258ff).

the objective pole and psychology is about the subjective one. This formula may actually be misleading, because phenomenological psychology is not about that which is subjective (or psychological) in the sense of everyday language: It is about the same objective structures as phenomenological philosophy, but with an emphasis on the acts of the subject. In this sense, we should perhaps rather say that linguistic methods such as commutation and grammaticality judgements are part of phenomenological psychology.[16]

If the observations made above are correct, one may wonder what the task of *triangulation*, as Zlatev uses the term, may be. When there are variations of different test results, on the side of third-person methods, and different descriptions in the self-generated protocols, on the side of the first-person methods, there is a common tendency to interpret the latter as simply being an epiphenomenon of the former. Even more seriously, when there are variations in the result of brain scanning or the like, on the side of third-person methods, and different descriptions in the self-reports, on the side of the first-person methods, the former will usually be seen as the cause of the latter (or at least relatively closer to the cause than the latter). In fact, of course, all we have are correlations, which can be interpreted in many other ways. But correlations, if they are not disproved in the long run, are already important discoveries. In at least one case they become essential: in order to compare human beings and other animals, for instance primates, we have no possibility of using first-person methods. In some cases (notably with humanly enculturated apes), second-person methods may be appropriate. Fortunately, these kinds are methods are increasing used, in spite of the difficulties, with non-human primates. But basically, we can only know animals from Third person methods. This really also applies to small children, before a certain age, which can be variable, whether the task requires the possession of language or some of semiotic means of communication will do. Therefore, we can only fully compare human beings and other animals, as well adults and infants, if we can correlate first-person, second-person and third-person methods.

Moreover, there is the possibility of doing "front-loaded phenomenology" (Gallagher & Zahavi, 2008, p. 38ff), perhaps more properly described as phenomenologically loaded experimental science, since it consists in taking the results of phenomenological descriptions of particular phenomena as a point of departure for setting up a test situations. However, the idea that such experiments can be used to confirm or disconfirm the phenomenological descriptions would have to be considered with circumspection.

3.2. The Nature of the Phenomenological Method

Phenomenology is a method of description.[17] As such, it should really be considered an empirical method. The phenomenological method is based on the fact that everything, which, in the normal course of events, is available to (at least human)

---

16. My own impression is that it is very difficult to draw the line between phenomenological philosophy and phenomenological psychology. There is no doubt at distinction to be made, but it needs phenomenological elucidation.

consciousness is present to this consciousness as something being outside of it. Consciousness is consciousness of something, and that something is outside of consciousness. This is what, in the Brentano-Husserl-tradition, is known as *intentionality*: the contents of consciousness are immanent to consciousness precisely as being outside of consciousness. Thus, we may describe a particular phase in the stream of consciousness as being an act in which something outside of consciousness becomes the subject of our preoccupation. In accomplishing such an act, we are directed to something outside of consciousness. In a much-used example, Husserl suggests we are involved with the perception of a cube. The cube, then, is given in consciousness as being outside of consciousness. The act itself is not conscious as such (in fact, it is in some way co-conscious, but not as the subject of an explicit intentionality).[18] However, it is possible to turn the beam of intentionality around, directing it to the acts themselves, in order to study the particular way in which they are organized within consciousness. Once we do that, we have shifted the terrain: we are into phenomenology. Indeed, this turning around to the intentional acts themselves is known as the *phenomenological reduction*.

There is a corollary to this: in order to treat the act as something having a value in itself, we have to ignore whether the object to which the act is directed exists or not. This is the so-called *epoché*, the suspension of belief. The epoché is not to be understood as a doubt as to the existence of the world outside the mind, not even as a methodological doubt, in the sense in which Descartes introduced this idea. The world is still there. It is merely momentarily bracketed. There is still another requirement for doing phenomenological analysis in Husserl's sense: we have to be directed to the general structures, rather than the individual character, of each given act. Such directedness to the general facts is known as the *eidetic reduction*. In order to attain the level of generality, we have to go through free variations in the imagination, also known as *ideation*, by means of which we vary the different properties of the act, in order to be able to determine which properties are necessary in the constellation, and which may be dispensed with. If, like Husserl, we start out from perception, we might want to vary the different ways of perceiving the cube. There are indeed many acts of perception that are still the perception of a cube, and even, more specifically, perception of this same cube. Most notably, of course, the cube may be seen from different sides, from different perspectives, only in part from a peep-hole, and so on. Suppose, however, that I see the cube as being something more, in particular: a die. I have then added a variation to the original pristine cube, which is now not simply a cube anymore (although, in this case, it still remains a cube at the same time). The question then becomes: is this still an act of perception. Husserl, clearly, would not

---

17. It is of course possible to consider phenomenology to consist of several methods, as suggested to me by Jordan Zlatev, since there are certainly several methodological operations involved (on which more will be said in the following). Although, as far as I know, Husserl never considered this possibility, it is not impossible that some of these operations may be accomplished independently. Without taking a stand on this issue in the present context, I have opted for following tradition in talking about *the* phenomenological method.

18. This observation, in itself, is an important result of Husserl's phenomenological studies, but we must ignore it in the following, in order not to complicate matters inordinately.

think so. I disagree (Cf. Sonesson, 1989). The fact that Husserl and I happen to draw different conclusions from the act of ideation in this particular case is important—similar things have certainly happened many times before in the history of the phenomenological movement, even, in fact, to Husserl himself when he repeated some of his own analyses. This does not mean that the results of phenomenological analyses can vary arbitrarily, as is often said about subjective approaches. On the contrary, all who have practiced phenomenology agree on the basic structures of phenomenological experience. But Husserl repeatedly invokes the necessity of a community of phenomenologists who would be able to corroborate existing phenomenological analyses.[19]

In spite of the terminology often used by Husserl (such as Wesenschau), phenomenological results do not present themselves in the form of any kind of revelation, given in a single instance. Rather, the phenomenological method supposes the accomplishment of an arduous work, which has to be done over and over again, in order to ascertain a reliable result. At least, this is, in actual practice, how Husserl went at the task: as can be seen in the numerous volumes of the *Husserliana*, published after Husserl's death, Husserl laboriously went through the same descriptions and variations over and over again, without being completely satisfied with the result. Indeed, as in all scientific endeavours, the result of the phenomenological method always remains provisional. This is what Husserl, with another rather misleading term, calls *Evidenz*. Peirce, of course, thought of this as the (potentially infinite) sequence of interpretants.[20] Some early phenomenologists, such as Aron Gurwitsch and Maurice Merleau-Ponty, went through some of Husserl's painstaking analyses once again, finding new facts about perception, the field of consciousness, and embodiment. I have myself taken upon myself the demanding task of elucidating the structure of the picture sign, finding in some cases that Husserl's own work was not meticulous enough (Sonesson, 1989). More recently, Thompson (2007) has applied himself to the charge of enhancing Husserl's analysis of mental images. There is no end to the work: but, as in all scientific endeavours, we always seem to get a little closer to the truth.[21]

Peirce, to be sure, was also into phenomenology. According to his definition, phenomenology is that particular branch of sciences which "ascertains and studies the kinds of elements universally present in the phenomenon, meaning by the

---

19. For more about this in recent secondary literature, cf. Zahavi & Gallagher (2008, ch. 2).
20. This, indeed, is a common interpretation of what Peirce says, but on close reading of Peirce's texts, is seems unavoidable to conclude that, to Peirce, truth is given beforehand, although we can only attain it by approximation: as he often explains, human beings and the world are of the same stock, so we will end up discovering the truth in the long run. A case in point is this paragraph in the Letters to William James (EP, 502f): "in other words our Reason is akin to the reason that governs the Universe; we must assume that or despair of finding out anything." There does not appear to be any such postulate of metaphysical historicism in Husserl. Now, also cf. note 57 in Stjernfelt (2007, p. 432) on Peirce entertaining rather a correspondence theory than a coherence theory of truth.
21. My description of the phenomenological method is based on what I have gathered from my own reading of Husserl's work, as well as my own attempt to use the method. A more principled, and recent, description of the phenomenological method can be found in cf. Gallagher & Zahavi (2008) and Thompson (2007, p. 267ff)

phenomenon whatever is present at any time to the mind in any way" (EP II, 259). Peirce himself claims to have taken the term from Hegel, but as has been pointed out by Stjernfelt (2007, p. 441, note 153), his usage of the term coincides with the period in which he was reading Husserl, and there are indeed obvious similarities between Peirce's and Husserl's usages, which are not found in Hegel's work.[22] Later on, Peirce was going to call the same study *phaneroscopy* and describe it as follows:

> a study which, supported by the direct observation of phanerons and generalizing its observations, signalizes several very broad classes of phanerons; describes the features of each; shows that although they are so inextricably mixed that no one can be isolated, yet it is manifest that their characters are quite disparate … (CP 1.286)

It would suffice to substitute the term *phenomenon* for *phaneron* to obtain a text that might be describing the phenomenological method according to Husserl (certainly the phenomenological and the eidetic reduction). Indeed, we have seen that Peirce himself used the term *phenomenon* in his earlier description of what he also at the time called phenomenology. Peirce's text, however, continues in the following way:

> … then proves, beyond question, that a certain very short list comprises all of these broadest categories of phanerons there are; and finally proceeds to the laborious and difficult task of enumerating the principal subdivisions of those categories. (CP 1.286)

Husserl, of course, would also expect some very broad categorizes to be established by this method. Nevertheless, it seems incompatible with his whole view of phenomenology to claim beforehand that a short list of such broad categories could be established. The difference between Husserl and Peirce becomes even more pronounced, when we realize that Peirce's short list will certainly be made up of triads comprising other triads, as well as some dyads and a few single terms.[23] This recursive triadic organization is a foregone conclusion of Peircean semiotics, which is prior to any phenomenological investigation, that is, is a priori, not because this has been established by free variation in of the imagination, but in the (French) ordinary language sense of being decided before any observation takes place. This is the first unjustified presupposition of Peircean phenomenology. But there is also another one, which concerns the content of the original triad, Firstness, Secondness, and Thirdness,

---

22. Stjernfelt (2007, p. 141f) quotes many examples of Peirce's definitions of phenomenology, which show clear similarities to Husserl. He also documents the mutually negative opinions the two scholars would have seemed to have on each other, clearly because none of the had really read the other. In the whole of this book, Stjernfelt (2007), and in particular on page 141ff, gives a lot of examples of the similarities between Husserl and Peirce. The precursor of such rapprochements is, interestingly, Spiegelberg (1956), otherwise know as the most authoritative historian of the phenomenological movement, in the strictly Husserlean sense. Here, we will however be concerned with only one similarity and how it turns out to lead on to dissimilarity.

23. Such as the representamen, which is Firstness lacking subdivisions, the object, which is Secondness, being divided into dyads, and the interpretant, which is Thirdness, being analyzed into different kinds of triads. However, the icon, in spite of being Firstness, is of three kinds: images, diagrams, and metaphors.

which are the meanings which are supposed to recur all through the hierarchy of triads. Thus, viewing Peirce's phenomenology from the end-point of Husserlean phenomenology, there are (at least) two postulates which have to be justified: that all categories come by threes (with the exceptions noted above), and the specific content of the three original categories.

At this point, Peirce's phaneroscopy could be considered to be one possible variant resulting from the Husserlean variation in the imagination—one that is not necessarily true, or which may be correct or not according to its particular instantiations, such as, just to mention the most obvious cases, Peirce's first, second, and third trichotomies.

However, before going on to a discussion of these trichotomies, I would like to take note of an important methodological addition to the phenomenological method which, as least in its explicit form, must be ascribed to Peirce: the use of diagrams, which, since they are icons, allow us to directly "see" the result of the variations, either in the graphs themselves, or with addition of some further presuppositions projected onto the graphs. We have all been using diagrams in this way, as even Husserl (for instance when discussing time consciousness) and Peirce did on rare occasions in the received texts (and as Peirce in addition did, more programmatically, in his existential graphs): my own table 1 above, as well as many others tables and figures in earlier works (e.g., Sonesson, 1989), Zlatev's table 1, reproduced above as table 2, and so forth. The importance of diagrams for resolving phenomenological problems, mentioned many times by Peirce, still remains the central insight, as well as the fundamental theme, of Stjernfelt's (2007) recent book.

3.3. The Trichotomies According to Peirce – and Calvino

The idea that all divisions of the (experienced) world come by threes is impossible to prove. However, it may be as impossible to disprove. Still, if we are not living in the kind of world parodied by Eco in *Foucault's Pendulum*, there is really very little chance that the world is actually made up in such a way. We are, of course, not talking about the way the world "really is" – but the way it appears to be to phenomenological description, and, at least according to the quotation above, it seems that also Peirce, when he was talking about his hierarchies of threesomes, was thinking about what was accessible to phenomenological observation (for, even admitting the existence of Peircean quasi-minds, Peirce recognizes that it is through ordinary human minds that we have privileged access to phenomena). The Peircean universe of discourse is regimented by the mystique of numbers, and to that extent, his work is part of a large-scale Western tradition with (at least partly imaginary) Oriental sources that construes the world (as we experience it) as being built on fixed quantitative relationships, which have an esoteric meaning. This is a world-conception testified from Antiquity to Giordono Bruno and Raymond Lullus (cf. Yates, 1964, 1966; Eco, 1995). Conceptions like these were for an appreciable amount of Occidental history part of the common sense world of at least some intellectual groups. This, however, does not show that such conceptions could be phenomenologically justified. As pointed out in section 1, the task of phenomenology is to reach beyond common sense.

It may, of course, be phenomenologically correct to say that, from some well-defined point of view, there are indeed three kinds of signs, with respect to the different relationships which may obtain between expression (representamen) and content (object and/or interpretant): that is, there are iconic, indexical, and symbolic signs. For a long time, I have indeed found this division intuitively satisfying, although I am still at a loss to say exactly from what point of view the variation in imagination must be made to obtain this result. Thus, one may feel that the distinction between expression and content being related by a mere regularity or by a normative imposition is too important to be conflated into one kind of relationship, the symbolic one (both being *habits* in the special Peircean sense of the term, which will be discussed below). But even if this division should turn out to be phenomenologically relevant, it does not follow that all other variations in the imagination must result in threefold divisions. Thus, for instance, the idea that the sign itself is made up of three parts is not phenomenologically justifiable as a matter of course. As I have argued elsewhere (Sonesson, 2007, in press a), the question whether something has two or three parts has no meaning before determining the domain for which the model is valid, as well as the criteria (the relevant properties) according to which the division is made. Since the domain of the Saussurean sign is that which is internal to the sign system, the content being all the time opposed to the "real world" it interprets, it would also be a triadic notion, to the extent that reality outside the sign system were included in the domain to be analyzed. As for the Peircean sign, it really comprises six instances, if all criteria of division are included, since there are two kinds of objects, and three kinds of interpretants, but only one kind of representamen.

Quite apart from the necessity of always making threefold divisions, there is the question of the content of each of the three categories. Firstness, Secondness, and Thirdness mean so much more than just being the first, the second, and the third category of an obligatory segmentation of the world into triads. Often, Peirce simply claims that Firstness is something that exists in itself, Secondness must be related to something else, and Thirdness requires a more complex relationship (either a relation between three things, or a relation between relations, or perhaps both at the same time). One of the more formal definitions of the three categories reads as follows:

> Firstness is the mode of being of that which is such as it is, positively and without any reference to anything else. Secondness is the mode of being of that which is such as it is, with respect to a second but regardless of any third. Thirdness is the mode of being of that which is such as it is, in bringing a second and third into relation to each other. (A Letter to Lady Welby, CP 8.328)

Firstness and Secondness could here almost be understood as a somewhat distorted equivalent of Husserl's (1913, Vol. II, part 1, 225ff) distinctions between independent and dependant parts, with the exception that there is no proviso for the difference between mutual and one-sided dependence (the same three-fold distinction made by Hjelmslev [1943], as Stjernfelt [2007, p. 167ff] judiciously remarks). This then raises the question what the business of Thirdness is. If it involves a relation between two terms, instead of only one term and a relation, as Secondness could perhaps be understood to be, or a relation between relations, why then should we not go on defining Fourthness, and so on? Of course, Peirce claimed that all relations beyond Thirdness could be dissolved into several relations, but Thirdness itself could not be so resolved. It is not clear whether this is indeed a phenomenological fact. Actually, this must, among other things, depend on what exactly is to be understood by Firstness, Secondness, and Thirdness. Thus, for instance, is there really no relationship in Firstness? When it is used to define a kind of sign, the icon, it must already be supposed to be part of a relationship, even before it is seen as a sign (namely the relation of similarity).[24] Indeed, Peirce repeatedly says that Firstness cannot be grasped as such. And what about Secondness? Is Secondness second, because it is made up of two things—in which case it would already be made up of three items, two things and a relation? Or should the second thing be conceived as a relation hooked up to an element, (Sonesson, in press b)? Thirdness, in a similar way, then would have to contain three hooks, one of which is already filled up with an element describing the nature of the relationship.

However, there are many places where Peirce imputes a much more concrete content to each of the categories. Since it is impossible to look at all the (only partly overlapping) descriptions of these categories offered all through Peirce's writings, a few instances pertaining to each category will have to do here: as I could not avoid to mention above, Firstness is the fleeting instant which can hardly be grasped in itself; indeed it is the present moment; it is quality; it is possibility ("except that possibility implies a relation to what exists, while universal Firstness is the mode of being of itself" [CP 1.531]); "freshness, life, freedom" (CP 1.302); "spontaneity" (CP 3.432); "indeterminacy" (CP 1.405); "agent," "beginning" (CP 1.361); it is also "immediate, new, initiative, original, spontaneous, and free, vivid and conscious"; before "all synthesis and all differentiation"; having "no unity and no parts" (CP 1. 357).

As for Secondness, is involves things such as "brute actions of one subject or substance on another" (CP 5.469); "the experience of an effort," "reaction," "resistance," and "opposition" (CP 8.330); "actuality," being "then and there" (CP 1.24; cf. "hæcceity" CP 1.405); "willing," "experience of perception," "existence" (CP. 1.532); "dependence" (CP 3.422); "patient" (CP 1.361); and so on.

Thirdness, finally, is "the mental or quasi-mental influence of one subject on another relatively to a third" (CP 5.469); "law" (CP 1.26); "habit" (CP 1.536); "general rule," "future" (CP 1.343); "cognition" (CP 1.536-537); "representation" (CP 5.66); "mediation" (CP 2.86-89 & CP 1.328); and so on.

After many decades of reading Peirce, I do find a kind of coherence in these categories, but it is certainly not in the form that can be subsumed by any ordinary definition characterizing necessary and sufficient conditions. And in spite of what is,

---

24. Thus, from the point of view of the sign, iconicity only starts being potentially interesting as an iconic ground, as we will see below, in section 3.2.

on the face of it, the diversity of the categories, their content is certainly much more specific than what is contained in the purely numerical definitions. Indeed, given these descriptions, Firstness, Secondness, and Thirdness certainly sound very much like what Vygotsky (1962) would have called chain-concepts, characteristic of small children (and what at the time were known as savages). Since Wittgenstein presented this type of vague concepts as based on *family resemblances*, and pointed out that they are spread all over ordinary language, it seems that they have been somewhat rehabilitated. Rosch conceived the idea of prototype-based categories, according to which a category is defined by a central example that seems to embody what is important to the category, with other members being at different distances from the prototype. In a number of experiments, Rosch showed this explanation model to make sense psychologically. One of the most interesting experiments involved placing objects on a spatial layout in relation to some object that was taken to be the prototype of the category. It would therefore seem that this notion is grounded by a combination of first-person and third-person methods.[25] Rosch and Mervis (1975) reflected on the relations between the prototype and Wittgenstein's family resemblances, arguing that the difference consists in the former being related to a central example, while the second lacks any such instance.[26]

At first, one may tend to see in the Peircean categories some kind of chain-concepts or family resemblances, but I think a few of the members of the chains can really be considered to make up the prototype, or perhaps really the ideal type, of the categories. This could be seen as a generalization of the claim, made over and over again by Peirce, that some instances of his categories are *degenerate* (a suggestive term in Peirce's work, the meaning of which we are left to guess at). The others, then, would be the prototypes or ideal types. In the case of Firstness, this central idea seems difficult to grasp, but it certainly has something to do with fleetingness or streaminess. Secondness is dominated by the idea of reaction/resistance. And law tends to be the most prominent element of Thirdness. It would be impossible to justify these contentions in the short space of this article. However, I think the following quotation from Peirce goes a long way in showing that (double-sided) resistance if the ideal type of Secondness:

> A door is slightly ajar. You try to open it. Something prevents. You put your shoulder against it, and experience a sense of effort and a sense of resistance. These are not two forms of consciousness; they are two aspects of one two-sided consciousness. It is inconceivable that there should be any effort without resistance, or any without a contrary effort. This double-sided consciousness is Secondness. (EP I, 268)

---

25. Indeed, it would seem to be front-loaded phenomenology *avant la lettre* (Cf. section 2.1).
26. Elsewhere, Rosch (1975) erroneously identifies her prototype concept with the Weberean ideal type. The incorrectness of this is shown by Sonesson (1989, p. 71f): whereas the prototype is defined by means of the central example of a category and includes as other members other items being at more or less great a distance from this central instance, an ideal type is an artificial creation, which is exaggerated in relation to reality and may contain contradictory properties, often projected onto time and/or space.

After half a century of familiarity with Peirce, I must say that the original threesome does start to make sense to me. The question, however, is how it compares to another conceivable triads, for instance the one created by the Italian novelist Italo Calvino (1983), for the purpose of a volume of short stories. All the stories are arranged into number combinations, which are recursive, just like the Peirecean triads, that is, they are applied to themselves (there is the Firstness of the Firstness of the Firstness, and so on). At the end of the book, Calvino offers a list of three categories: the first category involves visual experience, has often as object natural shapes: it could be termed description; the second category contains anthropological-cultural elements, language, meanings, signs: it is narrative; the third category incorporates speculative experience; it is about cosmos, time, infinity, the relation of self to the world, the dimensions of the senses; meditation. In the book, the part dedicated to the first category is about vacations, and as also Calvinean Firstness combines with Firstness of different degrees, the stories tell us about observations of waves, of naked breasts on the beach, and of the reflection of the sun in the sea; when it combines with Secondness, we have descriptions of animals; and in the combination with Thirdness, stars are described. The part involved with the second category as the dominant category plays out in the city, giving rise, in different combinations, to narratives of things taking place on the terrace, when shopping, and at the zoo. The stories comprised in the third part are determined predominantly by the third category, and deal with silences, which in the different subcategories involve trips, society life, as well as the universe and death.

Clearly, Calvino's categories only slightly overlap with those of Peirce. In fact, he certainly had no ambition of being a Peircean, since his connection to semiotics were actually on the side of the Greimas School. And although Calvino was a thinker not deprived of metaphysical aspirations, there is not much probability that he took his categories as seriously as Peirce did. In fact, Calvino's categories may appear easier to grasp than those of Peirce. When projected onto the stories, however, they give the same feeling of conceptual nebulae, which still somehow seem to make sense deep down.

3.4. The Onto-epistemology of the Phenomenological Method

It was claimed above that, from the point view of the kind of cognitive semiotics presented in this article, phenomenology is above all a method. But no method goes without a certain amount of ontological and epistemological presuppositions, and this is perhaps particularly true when it comes to phenomenology. So, in the following, I would like to sort out, at least in a preliminary way, which onto-epistemological consequences follow from applying the phenomenological method, and which do not. Since I do not know of anybody having tried to do precisely that, I hope I may be excused for touching the matter only briefly.

The primary presupposition of the phenomenological method is that consciousness exists. It is not, as Dennett would have it, just an illusion. Nevertheless, it is not entirely clear what the difference between having consciousness and being under the illusion of having consciousness is. The difference is not an obvious one like that between having a wart and only having the illusion of having one. If you think (or believe, etc.) that you have a consciousness, that would seem to be tantamount to having a consciousness. Probably the idea is a follows: If we have consciousness or not, this would not change anything else in the world. Since the world in question is obviously the world of our experience, it is of course extremely difficult to make any sense of the proposition that consciousness is an illusion. If this were true, however, the phenomenological method would be pointless. If the structures of the acts making up our stream of consciousness do not contribute to the way the world is present to us, there is nothing that could be investigated by the phenomenological method. For the same reason, there would be no object to be studied by what Dennett has called *heterophenomenology*. If, as Dennett claims, it is concerned to study, from a Third-person perspective, reports from the fictional domain resulting from the intentional stance, then, as Thompson (2007, p. 303ff) remarks, it would really be a Third-person method based on First-person methods, and it would not escape any of the problems pertaining to phenomenology proper. Moreover, as Gallager & Zahavi (2008, p. 18) add, there is no way you can interpret Third-person reports, without referring them to your own First-person experience, or else the interpretation must be based on unverified notion taken over from folk psychology. More importantly, however, such a study would be rather pointless, because it could not tell us anything about the way the world of our experience really is built up. To this final caveat Dennett would perhaps answer that to study the results of the intentional stance is as interesting as to investigate other works of fiction, from fairy tales to literary works. But this would mean that heterophenomenology is simply not playing the same game as phenomenology.

The primary ontological presupposition of the phenomenological method, then, is therefore that there is a particular domain of existence that could be called consciousness or mind. In this sense, mind cannot include what Peirce terms quasi-mind, if, as I believe probable, Peirce would be willing to accept the biosemioticians' idea of the cell having such a quasi-mind (which is at least implied by the idea of cells exchanging signs and the recourse to the Peircean sign definition in biosemiotics, e.g. Hoffmeyer, 2005). In an important sense of the terms, mind is not continuous with matter. It would seem to be impossible for a phenomenologist to accept, in Peirce's words, that "matter is effete mind, inveterate habits becoming physical laws" (CP 6.25), just as what could be taken to be the opposite, but in fact undistinguishable, doctrine that mind is simply matter. If Peirce's all-pervading conception of continuity (on which see Stjernfelt, 2007, pp. 1-48) should be taken to apply to the relation between mind and matter, in this sense, as suggested by Brier (2009), it would seem to render the phenomenological method impossible. We must of course take care (as Peirce should have done) to specify in what sense we claim a discontinuity between matter and mind. Short of being a full-fledged dualist, and even a creationist, one must necessarily admit that consciousness arises from matter, just as life has beforehand proceeded from inert nature. Similarly to life, consciousness can be seen as an emergent property. Nevertheless, I fail to understand Thompson's (2007, p. 221ff) argument that, since life, which seemed so mysterious to the vitalists, has been shown to result from the ordinary laws of chemistry, we should accept that mind is not any more mysterious, although there is, at present, no such straight-forward explanation for its emergence. No matter the chemistry, on the contrary, life remains a big hiatus from a phenomenological point of view; and so does mind.

The second important presupposition of the phenomenological method must rather be assigned to the side of epistemology (and, in fact, also to psychology). It is one thing to have consciousness, and it is another to be able to take cognizance of the fact of having consciousness. If our consciousness is normally made up of acts directed to the outside world, and, in particular, if this directedness to the outside world is an intrinsic property of consciousness (that is, intentionality), the acts themselves could easily be outside of our awareness—and perhaps unavoidably so. But the phenomenological method supposes it to be possible for the subject to turn around and observe its own acts. The subject, then, would have to be able to make normally unconscious, preconscious, or semi-conscious parts of mind into objects of awareness.[27] More specifically, the very acts by means of which contents normally become conscious have to be made available to consciousness – precisely as being the acts in which certain contents become subject of our awareness. This is an epistemological issue, to the extent that it asks whether the phenomenological method is qualified for the task. However, it is also a problem of (phenomenological) psychology, because it requires real human beings of flesh and blood and, more importantly in the circumstances, possessing the kind of minds which are empirically known to exist, to be able to accomplish feats of a nature corresponding to operations operating on their own operations, while being aware of the contents of the original operations. And if this formulation causes your mind to twist, you may be able to appreciate the difficulty involved.[28]

Not surprisingly, no tick or bat, nor any other animal has been known to use the phenomenological method. No cell could be expected to do it on its own. The phenomenological method is difficult to use, also for human beings, as testified most clearly be the posthumous writings of Husserl (simply because nobody else seems to have set down all their provisional phenomenological analyses in writing). It is, as all scientific methods, eminently fallible, to adopt a Peircean expression. However, even if we suppose some kind of *qualia* to accompany the simple life of the senses characteristic of bats (and perhaps even ticks), it would be very surprising to learn that they also had some *meta-qualia*, that is, some experience of their own experience. For all we know, chimpanzees, bonobos, and other primates may well be great

---

27. The term *preconscious* should not be understood in the Freudian sense, but almost literally, as that which is not a subject of awareness at the moment (and/or habitually). It goes without saying that the term *unconscious* should not be understood in a Freudian, Lakoffian, or any other mystical sense. Nor should it be understood as equivalent to physiology and/or what is in present day jargon known as the subpersonal level.

28. Although this is subject to further investigation, this seems to me to be a tougher requirement than the kind of meta-operational position while abstracting from the contents of the individuals acts, which Piagetean psychology supposes to be a stage we all attain during development.

phenomenologists, simply being unable to communicate their discoveries to us, even by means of the artificial sign systems that we have taught them. More probably, however, they are too busy, as most human beings are, living in the acts of immediate experience of the world. After all, the phenomenological method is a late-comer even in human history.

Without pretending to have spelled out all the onto-epistemological presuppositions of the phenomenological method, I would now like to say something about what it not implied by the method, in spite of these implications often being taken for granted. First of all, the phenomenological method does not suppose any solipsism. A solipsist may of course go through all the operations of the phenomenological method, but he would then only learn something about himself. His situation would not be unlike that of Dennett's heterophenomenologist. The first problem, which Husserl tackled in his *Cartesianische Meditationen* (Husserl, 1950), which should perhaps more properly be called the Anti-Cartesian Mediations, was the fact that other humans beings are not present to consciousness simply as contents of consciousness (or as automata, as Descartes suggested, looking on other human beings from afar): They are there as other subjects being able to make me the object of their intentions in the same way that I make them the subject of mine. As an anti-psychologist, in the sense of his time, Husserl was of course also concerned to show that some structures, which were not (exclusively) mental, such as, in a famous example, geometry, had some existence outside of individual minds. Finally, he even ended up claiming that the perception of our common perceptual world must be grounded, not only, as supposed in the early texts, on the possibility of going round those objects which we do not now see in full, to contemplate them from other angles of vision (Ich kann immer weiter), but in the conviction that these objects are (at least potentially) experienced by other human beings from other points of view (cf. Zahavi, 2003).

These are problems encountered by phenomenology, and they have certainly not been fully resolved as yet. Those who point out that these are problems for the method, however, tend to be people who reject the method. They fail to inform us (or even becoming aware of the fact) that, outside of the phenomenological method, there is no way of even starting to resolve these problems, because, without the phenomenological method (or some precursor to it), there is no way you can even be aware of these problems.

None of this has anything to do with Husserl's famous turn to transcendental idealism. That you can use the phenomenological method without being a transcendental idealist is shown by the fact that those few thinkers who remained close to Husserl's methodological precepts, such as, to pick two rather different examples, Ingarden and Gurwisch, both rejected this metaphysical reinterpretation on the part of Husserl—without any ensuing incoherence.

No ontological idealism is implied. The world is there to us only *through* consciousness, but not *in* consciousness.

3.5. From Historical Structures to the World of the Natural Sciences

So far, I have been concerned with the onto-epistemological presuppositions of the phenomenological method as such (without any claim to exhaustivity). But there are also, I believe, provisional results of the method, some stemming from Husserl himself and his few close followers, some from the only other phenomenologically inspired tradition of semiotics I know of, the Prague school, and some resulting from my own work.

Cognitive science is, it seems, a rather ahistorical tradition, although it has, in recent times, taken on some evolutionary issues. Both phenomenology and semiotics, on the other hand, have come under fire from neighboring traditions, because of a purported lack of interest in diachrony, whether this is to be understood on the time-scale of evolution, or that of human history. Within semiotics, the issue is complicated by the fact that one of the founding-fathers, Peirce, did manifest some interest in evolution in the Darwinian sense of the term, while the other, Saussure, is famous for insisting, after several centuries of linguistics being completely dedicated to historical studies (although this was the kind of studies in which he made his name in his own time), on the importance of studying the structure of meaning systems synchronically. A lot could be said in order to refute the claim of phenomenology and semiotics ignoring the diachronic dimension, but the historical fact of the Prague school, based on a double phenomenological and semiotic heritage, is sufficient to deny the accusation in both cases. In the most extreme version of its claims, the Prague school would say that there is always already diachrony in synchrony, and that diachrony can only be applied to synchrony (i.e., to entire structures). Indeed, in the terms of the title of an excellent book about the Prague school (Galen, 1985), the structures as such are historical. It is also, unfortunately, an historical fact that the Prague school ceased its intellectual work in the forties of the last century, and that very few of those involved with semiotics since then have seriously taken on this heritage.

Evolution does, however, appear to be a trickier issue. It may seem that history could still be conceived from within phenomenology, in particular with a rather common hermeneutic extension, as the subject taking on the tradition handed down to it from earlier history. If anything, Darwinian evolution is outside experience. Evolution does not have any obvious subject. Indeed, there can be not phenomenology of evolution. This, however, is no more strange or disturbing than the fact that phenomenological analyses cannot take over the function of brain-scanning. Phenomenology can bring semi-conscious, pre-conscious and unconscious parts of consciousness into awareness. Husserl brings this transformation as far as kinesthetic experience. Biological evolution, however, is something which only exists at the subpersonal level (Bermúdez, 2005), which, as the terms clearly says, is a level on which there is as yet no subject. Furthermore, it also starts at a point in diachrony in which subjects did not yet exist, and, in any case, at a moment in time for which no subject living today could possible by able to reach, however much he or she tried, any semi-, pre- or unconscious experiences of its own. This fact does not serve to highlight any defects of the phenomenological method. On the contrary, the reflections on the phenomenological method offered in this paper demonstrate that any knowledge we may have about evolution, and about nature in general, must be indirect, mediated by a third-person perspective.

Husserl addresses both our present concerns in *Krisis* (Husserl, 1954), the diachronic one, unfortunately, only in terms of human history, but the relationship of phenomenology to the natural sciences on a broader scale. To begin with, this is the first time Husserl presents phenomenology as a method that is historically situated. It does not just happen to be invented at the turn of the 19th century by Husserl himself, following ideas about an empirical or descriptive psychology formulated by Franz Brentano (1924-28), then being at least in part rediscovered by Peirce. The invention of phenomenology, in Husserl's view, was historically necessary at this very moment (though not in the sense of any Hegelian historic predetermination). After the invention of the natural sciences a few centuries earlier, and the generalization of a rationalist attitude to all kinds of problems, including those characteristic of the human and social sciences, which took place during the Enlightenment, science and rationality had made ever greater strides. This process produced during Husserl's own time a crisis in the relationship between the sciences and the Lifeworld, that is, the *world taken for granted,* as Husserl's close follower Alfred Schütz was later to call it: the world in which ordinary human beings, including the scientists *as* human beings, stake out their life.

So far, Husserl's quandary seem very close to the dialectics of the Enlightenment diagnosed by Horkheimer & Adorno (1947), according to which reason run wild has ended up taking over all of experience, depriving ordinary life of its magic, which, in the end, results in reason being instrumentalized into the industrial extermination of the Jews during the Second World War. As Michel Foucault (1994) was to add much later, very much in the spirit of the Frankfurt school, this reason run wild was involved in a much bigger project, transforming human beings into docile subjects completely regimented by different disciplinary systems, which, apart from prisons and mental hospitals, include the scientific domains themselves. Like the Frankfurt school, Foucault apparently saw this a completely negative development, which could however not be changed. Husserl's stand is quite different. Like the poet John Milton, who set out to justify the way's of God to man, Husserl is out to champion science in the eyes of human beings living in the Lifeworld. It is precisely for this reason that phenomenology is needed. The phenomenological method discovers the rational structures that are at the foundation of the Lifeworld.

In spite of that, (natural) scientists will easily get the impression that Husserl sets out to relativize the achievements of science. There is a danger of forgetting that the only real world of our experience is that of the Lifeworld, and that even the scientist, when he makes his experiments, is himself located in that world. So are his instruments: If the scientist uses a particle accelerator in order to study the internal structure of particles, he cannot treat the accelerator itself as an object of study, that is, consider it from the point of view of the particles it contains. Both the scientist and his apparatus must be firmly entrenched in the Lifeworld, if he is to be able to bring about a study of the scientific object of his domain. It is only from the point of view of the Lifeworld, that the scientific world can be conceived. This idea was later generalized by Alfred Schütz (1967), for whom the scientific world, together with the world of dreams, of phantasm, of play, of art, and of religion, were all "finite provinces of meaning"(Schütz, p. 229), accessible only taking your point of departure in the Lifeworld.

It would seem to follow from this that also the scientific world is a world of fantasy, if not phantasms. I think this is, in a way, a correct interpretation. But the scientific world is not a world of unregimented fantasy (like in the anarchist conception of Feyerabend). In Husserl's view, science makes progress, so it is certainly on its way to the truth, however far the latter may still be distant from us. Although a self-described phenomenologist, Barry Smith (Smith, 1999; Smith & Varzi, 1999) invokes ecological terminology to describe the Lifeworld as the niche in which human beings stake out their life. It is found on a *mesoscopic* level, in between the microscopic and the macroscopic levels described by physics, but it is real in the same sense as the latter two. According to Smith, this is James Gibson's view, which he opposes to that of Husserl, for whom only the Lifeworld, not that of the physical sciences, is real. This is not how I read Husserl; and even if it should turn out that I am wrong in my interpretation of Husserl, I should still prefer to see the relationship otherwise (cf. Sonesson, 2001). To take a familiar though perhaps somewhat old-fashioned example, physics may describe light as being at the same time a series of waves and a conjunction of bodies. This is nonsense to common sense, and rightly so: for, clearly, this must mean that light is really some third kind of thing, which happens to share some properties with the common sense objects called waves and bodies. So, the language that physics uses to describe the physical world is approximate and metaphoric. Similarly, in the 20th century, the atom was long conceived on the model of a planetary system in miniature, until it was suggested that electrons should rather be seen as amounts of energy transferred from different positions around the nucleus. These are two metaphors for the basic structure of the physical world, the first of which could for a long time account for the data known, later on being judged insufficient for the same reason. In this sense, I think, Husserl talks about the different systems of explanation of the natural sciences as being some kind of *Ideenkleid* cast over reality, which we should not mistake for reality itself. This does not mean that the world that this clothing of ideas tries to approach is not real. But the Lifeworld is the only world to which we have direct access and which may be described in its own language.

It is in fact unclear to me why Smith construes an opposition between Husserl and Gibson. As far as I understand (and as I have been suggesting at least since Sonesson, 1989), Gibson's psychology of perception is really the first attempt at naturalizing phenomenology. In fact, James Gibson's wife, Eleanor Gibson (Gibson, 1969; Gibson & Pick, 2000) who was more specifically involved with learning, describes the Gibsonean theory (that of them both, in fact), as being based on differentiation, rather than enrichment. This means that, in a sense, everything we may know is already present to perception, but it has to be discovered in the whole of which it is a part and put into the proper perspective. In *Pictorial Concepts* (Sonesson, 1989, p. 316ff) I illustrated this by what I called the wine taster's code. In the beginning, all wines taste alike. With time, we may learn to discover more nuances of taste, and in the end these differences seem so great that we have difficulty understanding that they at first appeared slight. But the tastes and their differences were there all along. This is of course a process taking place entirely in the Lifeworld. We may come closer to the kind of differentiation characteristic of the natural sciences, if we manage to account for the differences of taste separating the chemical components to which the taste differences are due. However, here we are still very close to the surface of the Lifeworld. When developing theories of the microscopic and macroscopic world, we normally only have access to a few facts visible from the standpoint of the mesoscopic world, to use the term proposed by Smith, most of the time only be means of some special kind of apparatus situated in the mesoscopic world. No full account can be made of what happens in both these worlds foreign to us, so we have to connect a few facts and invent a name for them, whether it is mitochondria or black wholes. These facts, as far as they go, are real enough. But their names and whatever they suggest in ordinary language are pure fantasy.

4. Stepping up to Husserl's Lectern: Phenomenology Within the Tradition of the Enlightenment

When Husserl wrote his *Krisis* in the 1930s (Husserl, 1954), it was still possible to feel proud of being a European. There is no reason to think that Husserl and his contemporaries ignored the fact that rational techniques had been put to bad ends submitting other countries and transforming them into colonies forming part of one of the many European empires of different scales still having course at the time. Although Husserl himself was to become a victim of the *Third Reich*, he did not necessarily realize that rational procedures could be used in the pursuance of irrational ends such as the extermination of all people of a certain stripe, as most notoriously exemplified by the cases of the genocides of the Jews and Gypsies; he was more certain to know about the irrational ends. No one, at the time, probably had any inkling of the idea that all of European civilization, at least from early Modern time onwards, could be conceived as the history of an increasing regimentation of the individual for the profit of the general society and the authorities that be, as was more recently suggested, in rather different terms, by Norbert Elias (1939) and Michel Foucault (1994). When Horkheimer (1967) wrote about these issues, without the help of his friend Adorno, he separated instrumental reason from reason *tout court*. Foucault, of course, being a man of his time, did not believe in any reason at all, except as a negative factor of control (in spite of using it, rather successfully, in his own work). He did believe in the (momentary) success of action in the street. It is not clear, however, whether a systematically organized genocide, such as that accomplished by the Nazis, is necessarily worse than a genocide without any clear systematic plan, such as appear to have been the more recent genocides realized in former Yugoslavia and in Rwanda. Postmodern critics, who point to the fact that rational techniques, whose origin they tend to place in the Enlightenment, have been put to all kinds of unethical ends, forget that, without the Enlightenment, there would be no one around to criticize these techniques, because, without the Enlightenment, these killings would be accepted as a matter of course, as one ethnic group defending itself or expanding its territory at the expense of another group.

When Husserl was proud of being part of the Enlightenment tradition, and of putting phenomenology at its service, he was, I believe, thinking of quite a different aspect of this tradition, and one which is, I believe, more essential. In his famous essay, Kant described the Enlightenment as the process by means of which human beings manage to liberate themselves from their self-occasioned state of minority (Kant, 1974). No doubt, emerging from a century of coffee-houses and political clubs (cf. Sennett, 1977), even the solitary Kant in Konigsberg may have been harbouring thoughts about politics (as he certainly did, in our sense of the term, in the case of the ever unfashionable theme of peace), but there can be no doubt that he had a much wider cultural range of change in view. Unfortunately, the liberation of human beings from the state of minority, as all other Enlightenment precepts, has since then been understood too exclusively in a limited political sense (as most clearly is the case in Habermas' 1995 work); and since it has long since been clear that representative democracy is representative of nothing (as has recently, at least in Sweden, been realized even by some of the elected representatives themselves, at the occasion of the passing of the laws limiting personal integrity on the Internet), the contribution of the Enlightenment has not been appreciated to its due value.[29]

Thus, for instance, concepts such as the bourgeois public sphere (Habermas, 1962) misconstrue history. No doubt the creation of the public sphere was at least partly facilitated by the advancement of the Third Estate during earlier centuries, the vanguard of which was at the time the class of wealthy capitalists, simply because they had the economical means to out-weight the traditional power of the landed classes. However, until the end of the French Revolution, the Third Estate comprised all social groups that were neither nobles nor clergy. As for the consequences of the public sphere, they were, and still are, (in part) beneficial, as far they go, to all of the Third Estate, which now is more or less co-extensive with humankind, although, now as then, the access to the sphere is regimented by hard social facts deriving from economical and other kinds of political power. However, if the public sphere only comprises the debates in the Parliament, the writings in the daily papers, and the discussions taking place in the coffee-houses (Sennett, 1977), nowadays no doubt substituted by blogs, Facebook pages, and Twitter contributions, then I am not sure

---

29. Of course, there is only a limited political sense to us, not to Kant. To him, as to many important figures of the Enlightenment, representative democracy may have seemed a fantastic (even Utopian) idea (but to others, who were rather elitist, the idea would be rather frightening), which would open up an arena of general rational discussion. For complex reasons (the age-old power games, the development of the media, and so on), things turned out differently.

that the Enlightenment was really worth-while. If we understand the Enlightenment as Kant did, however, and like the principle actors involved in the process known by that name, such as Voltaire and Diderot, and, later on, Destutt de Tracy, Cabanis, Degérando, and so on, then the whole point of the public sphere was the creation of a space where rational discussion was possible, far from the prejudices created by all sorts of traditional opinion, whether is was based on religious, political, or any other point of view (cf. Gusdorf, 1978; Rosenfeld, 2001). In this sense, the public sphere, rare at it is in the real world, is a space where rational discussion comes to its fore.

And in this sense, it is something to defend, because, as Husserl intimated, it has only emerged once is history, in Europe, and although it is a tradition that has had a lot of difficulty surviving also in Europe, it seems clear that elsewhere, something of the kind was apparently not even possible to conceive. Perhaps there is no reason to be proud of being European: Rather, we who live in Europe should be happy that something fantastic happened on the way from the zoo, which, so far, did not occur in other cultures. However frail its existence, the emergence of rationality in Europe is another European miracle, perhaps not entirely unconnected to, but not obviously explainable from, the more historically researched theme of the origin and development of Capitalism in Europe (Jones, 2003). This is why Husserl, at the time, talked about a crisis in the European sciences. From this point of view, already at the time and certainly now, the middle classes of Mexico, Turkey, and so on, are as European as we are. When Arab immigrants in European countries (no doubt rightfully) complain about their maltreatment opposing Islamic to Christian values, they are completely mistaken (unless they happen to be living in the United States, where the idea of the crusades still seems to be alive): what we have, even those of us who don't know anything about history, is really Enlightenment values.

This is important, because what the Enlightenment is really about is creating the possibility for rational discussion about anything whatsoever, with no limitation set by religious, political, or other ideological conceptions. Contrary to what Habermas suggests, these kinds of debates are not particularly geared to politics (in the present-time sense of party consensus), for the kind of rationality which may be applied to political life is very remote from the actual life of the people, as is clear from the election results even in countries where corruption and violence do not play any important part, and when rational approaches to politics have been tried out, as in the former Soviet Union and similar countries, the result has not been particularly impressive. At the other end of the ideological spectrum, most of the so-called science of economics has been dominated by the ideal type of rational man, which, within economics itself has only in the last few years come in for criticism, by for instance the recent Nobel prize winner Paul Krugman, although for everybody outside of economics, the capitalist idea of the market regimented by free and rational choice cannot but appear, even before the recent depression, as a paradox without any hope of (rational) resolution. I do not say any of these things in order to criticize politics and economics, but only to rescue rationality from their spell.

The scene where the issues of rationality are really played out, Husserl and Peirce would agree, is that of the sciences. It was, in the end, Husserl's conviction that the natural sciences needed phenomenology to be fully rational. Though Husserl never said so, as far as I know, a similar phenomenological grounding is also needed for the human and social sciences—that is, the (primary) semiotic sciences. But phenomenology, in this sense, is merely an outgrowth of a rational—Enlightenment—view of life. The Enlightenment is a process that has hardly begun.

Rationality, in this sense, is a process that is always on the way to something better. It takes place within the Peircean community of researchers, which, in this respect, is equivalent to the Husserlean *Ich-Gemeinschaft*. It is a rationality in search of the ever escaping *Evidenz*, or, according to a rather apocryphal interpretation of Peirce, of the unattainable final interpretant. In this sublunary world of our existence, rationality precisely means this: ever to strive for better knowledge, while having the criteria to determine that what we know now is at least somewhat closer to the truth than what we knew yesterday. In those sciences of which I have a close acquaintance, such as linguistics and semiotics, this is certainly not a fact at the present historical moment. This is exactly why cognitive semiotics needs to be phenomenologically conceived.

Acknowledgements

Although I have long been on the way to applying some kind of meta-theoretical considerations on what I have been doing for so long, in particular since the start of my present work in the *Centre for Cognitive Semiotics* (A programme financed at Lund University since the beginning of 2009 by the Tercentenary Found of the Swedish National Bank), many of the issues to which I direct my attention in the present essay have been prompted by questions posed to me by Søren Brier. However, if this essay is, in particular at the beginning, uncharacteristically centred on my own persona, in the sense of being very much concerned about what I have written before, as an anonymous reviewer pointed out, this is due to the challenge set by Brier. This essay also incorporates parts of earlier articles, on the occasion of which I received useful comments from Jordan Zlatev and Chris Sinha. For valuable comments on the present article, I want to thank Jordan Zlatev and an anonymous reviewer.

References

Bermúdez, J. L. (2005). *Philosophy of psychology: A contemporary introduction*. New York: Routledge
Bouissac, P. (1999). *Semiotics and the science of memory*. Paper presented at the conference on Semiotics and the European Heritage, Dresden, February 1999.
Brentano, F. (1924-1928). *Psychologie vom empirischen Standpunkt*. Leipzig: F. Meiner.
Bühler, K. (1934). *Sprachtheorie: die Darstellungsfunktion der Sprache*. Jena: Gustav Fischer
Calvino, I. (1983) *Palomar,* Torino: Einaudi.
Cassirer, E., (1942) *Zur Logik der Kulturwissenschaften*. Göteborg: Elanders.
Dennett, D. C. (1987). *The intentional stance*. Cambridge, MA: The MIT Press.
Dreyfus, H. L. (1992). *What computers still can't do: A critique of artificial reason*. Cambridge, MA: The MIT Press.
Colapietro, Vincent (1989). *Peirce's approach to the self: A semiotic perspective on human subjectivity*. New York: State University of New York Press.

Coseriu, E. (1978) *Teoría del lengauage y lingüística general*. Madrid: Gredos.

Daddesio, T. C. (1995). *Of minds and symbols*. Berlin: Mouton de Gruyter.

Deacon, T. (1997) *The symbolic species: The co-evolution of language and the brain*. New York: Norton.

Donald, M. (1991). *Origins of the modern mind*. Cambridge, MA: Harvard University Press.

Donald, M.(2001). *A mind so rare*. New York: Norton.

Eco, U. (1988). *De los espejos y otros ensayos* (C. Moyana, Trans.). Barcelona. Lumen. [Spanish translation of *Suglie specchi e altri saggi*. Milan: Fabri 1985].

Eco, U. (1995). *The search for the perfect language*. Oxford: Blackwell.

Elias, N. (1939). *Über den Prozess der Zivilisation: soziogenetische und psychogenetische Untersuchungen*. Basel: Haus zum Falken.

Espe, H. (1983a). Empirische Analyse visueller Zeichen: der Einfluss der Belichtungsdauer bei der Vergrössung auf die affektive Bedeutung von Schwarz-Weiss-Fotografien. In M. Krampen (Ed.), *Visuelle Kommunikation und/ oder verbale Kommunikation* (pp. 92-121). Hildesheim: Olms Verlag/Hochschule der Künste.

Espe, H. (1983b). Realism and some semiotic functions of photographs. In T. Borbé (Ed.), *Semiotics unfolding. Proceedings of the second congress of the International Association for Semiotic Studies, Vienna 1979* (vol. III, pp. 1435-1442). Berlin: Mouton.

Ferris, M. (2002). *Historia de la hermenéutica* (A. P. Cortés, Trans). Mexico City: Siglo XXI editores. (Italian original: *Storia dell'ermeneutica*. Milan: Bompiani, 1988).

Fodor, J. (1987). *Psychosemantics*. Cambridge, MA: The MIT Press.

Foucault, Michel (1994). *Dits et écrits*. Paris: Gallimard

Galan, F. (1985). *Historic structures: The Prague School project, 1928-1946*. London: Croom Helm

Gallagher, S. (2005). *How the body shapes the mind*. Oxford: Clarendon Press.

Gallagher, S., & Zahavi, D. (2008) *The phenomenological mind*. New York: Routledge.

Garfinkel, H. (1967). *Studies in ethnomethodology*. Englewood Cliffs, NJ: Prentice-Hall

Gibson, E. (1969). *Principles of perceptual learning and development*. Englewood Cliffs, NJ: Prentice-Hall.

Gibson, E., & Pick, A. (2000). *An ecological approach to perceptual learning and development*. Oxford: Oxford University Press.

Gibson, J. J. (1966). *The senses considered as perceptual systems*. Boston: Houghton Mifflin Co.

Gibson, J. J. (1979). *The ecological approach to visual perception*. Boston: Houghton Mifflin Co.

Gibson, J. J. (1982). *Reasons for realism* (E. Reed & R. Jones, Eds.). Hillsdale, NJ: Lawrence Erlbaum.

Goodman, N. (1968). *Languages of art*. London: Oxford University Press

Greimas, A. J. (1970). *Du sens*. Paris: Seuil

Greimas, A. J. & Courtés, J. (1979). *Sémiotique: Dictionnaire raisonné de la théorie du langage*. Paris: Hachette.

Gurwitsch, A. (1957). *Théorie du champ de la conscience*. Bruges: Desclée de Brouver.

Gurwitsch, A. (1974) *Phenomenology and the Theory of Science*. Evanston: Northwestern University Press.

Gusdorf, G. (1978). *Les sciences humaines et la pensée occidentale. 8, La conscience révolutionnaire: les idéologues*. Paris: Payot

Habermas, J. (1962). *Strukturwandel der Öffentlichkeit: Untersuchungen zu einer Kategorie der bürgerlichen Gesellschaft*. Neuwied: Politika.

Habermas, J. (1995). *Theorie des kommunikativen Handelns*. Frankfurt: Suhrkamp

Haser, V. (2005). *Metaphor, metonymy, and experientialist philosophy: Challenging cognitive semantics*. Berlin: Mouton de Gruyter.

Hjelmslev, L. (1943). *Omkring spogteorins grundlæggelse*. Copenhagen: Akademisk Forlag.

Hjelmslev, L. (1959). *Essais linguistiques*. Copenhagen: Travaux du cercle linguistique de Copenhague.

Hjelmslev, L. (1973). *Essais linguistiques II*. Paris: Minuit.

Hoffmeyer, J. (2005). *Biosemiotik: en afhandling om livets tegn og tegnenes liv*. Charlottenlund: Ries. (English translation: *Biosemiotics: An Examination into the Signs of Life and the Life of Signs*. The University of Scranton Press of Scranton, Pennsylvania, 2008)

Horkheimer, M. (1967). *Zur Kritik der instrumentellen Vernunft: aus den Vorträgen und Aufzeichnungen seit Kriegsende*. Frankfurt am Main: S. Fischer.

Horkheimer, M. & Adorno, T. (1947). *Dialektik der Aufklärung*. Amsterdam: Querido

Husserl, E. (1913). *Logische Untersuchungen*. Tübingen: Max Niemeyer.

Husserl, E. (1950). *Cartesianische Meditationen und Pariser Vorträge. Husserliana: Gesammelte Werke. Bd 1*. Haag: Nijhoff.

Husserl, E. (1954). *Die Krisis der europäischen Wissenschaften und die transzendentale Phänomenologie: eine Einleitung in die phänomenologische Philosophie. Husserliana: Gesammelte Werke. VI, 2. Aufl.* Haag: Nijhoff

Husserl, E. (1962). *Phänomenologische Psychologie. Husserliana: Gesammelte Werke* IX. The Hague: Nijhoff.

Husserl, E. (1980). *Phantasie, Bildbewusstsein, Erinnerung. Husserliana: Gesammelte Werke* XXIII. The Hague: Nijhoff.

Itkonen, E. (1978). *Grammatical theory and metascience: A critical investigation into the methodological and philosophical foundations of 'autonomous' linguistics*. Amsterdam: Benjamin

Ivins, W. (1953) *Prints and visual communication*. Cambridge, MA: Harvard University Press.

Jones, E. (2003) *The European miracle: environments, economies, and geopolitics in the history of Europe and Asia* (3rd ed.). Cambridge: Cambridge University Press.

Kohák, E. (1989) Preface. In E. Kohák (Ed.), *Jan Patočka. Philosophy and selected writings* (pp. ix-xiv). Chicago: University of Chicago Press;

Kant, E. (1974). Beantworting auf der Frage: Was ist Aufklärung? In *Was ist Aufklärung?* Bahr, E. (Hrsg.), 8-16. Stuttgart: Reclam.(Originally published in 1783)

Krampen, M. (1991). *Children's drawings: Iconic coding of the environment*. New York: Plenum Press

Lakoff, G. & Johnson, M. (1999). *Philosophy in the flesh: The embodied mind and its challenge to Western thought*. New York: Basic Books.

Lindekens, R. (1971) *Eléments pour une sémiotique de la photographie*. Paris: Didier/Aimav.

Lotman, J. M., Uspenskij, B. A., Ivanov, V. V., Toporov, V. N. & Pjatigorski, A. M. (1975). *Thesis on the semiotic study of culture*. Lisse: The Peter de Ridder Press.

Merleau-Ponty, M. (1972) *La structure du comportement*. Paris: PUF. 7ème edition 1972. (Originally published in 1942)

Nelson, K. (1996). *Language in cognitive development: emergence of the mediated mind*. Cambridge: Cambridge University Press.

Parmentier, E.. (1985) Signs's place in medias res: Peirce's concept of semiotic mediation. In Mertz, E., & Parmentier, R. (Eds.), *Semiotic mediation: Sociocultural and psychological perspectives* (pp. 23- 48). Orlando: Academic Press.

Peirce, C. S. (1931–1935, & 1958). *The collected papers of Charles Sanders Peirce*. Vols. I–VI [C. Hartshorne & P. Weiss, Eds., 1931–1935], Vols. VII–VIII [A. W. Burks, Ed., 1958]. Cambridge, MA: Harvard University Press. (Citations use the common form: CP vol.paragraph).

Peirce, C. S. (1998). *The essential Peirce* (vols. I-II, Peirce Edition Project, Ed.). Bloomington: Indiana University Press (Citations use the form: EP Vol., Page).

Piaget, J. (1945). *La formation du symbole chez l'enfant*. Neuchatel: Delachaux & Niestlé. Third edition 1967.

Piaget, J. (1967). *La psychologie de l'intelligence*. Paris: Armand Colin.

Piaget, J. (1970). *Epistémologie des sciences de l'homme*. Paris: Gallimard.

Prieto, L. (1975a). *Pertinence et pratique. Essai de sémiologie*. Paris: Minuit.

Prieto, L. (1975b). *Essai de linguistique et sémiologie générale*. Genève: Droz.

Radnitzky, G. (1970). *Contemporary schools of metascience: I. Anglo-Saxon schools of metascience. II. Continental schools of metascience* (2nd rev. ed.). Gotheburg: Akademiförl.

Ricœur, P. (1975). *La métaphore vive*. Paris: Seuil

Rosch, E. (1975). Cognitive representations of semantic categories. *Journal of experimental psychology: General, 104* (3), 192-233.

Rosch, E. & Mervis, C. (1975). Family resemblances: Studies in the internal structure of categories. *Cognitive Psychology, 7*, 573-605.

Rosenfeld, S. (2001). *A revolution in language: The problem of signs in late eighteenth-century France*. Stanford, CA: Stanford University Press

Saussure, F. (1973). *Cours de linguistique générale*(Edition critique par Tullio de Mauro). Paris: Payot.

Saussure, F. (1974). *Cours de linguistique générale. Fasc. 4*. (Edition critique par Rudolf Engler). Wiesbaden: Harrassowitz.

Schütz, A. (1967). *Collected papers I.* The Hague: Martinus Nijhoff.

Searle, J. (1997). *The mystery of consciousness*. London: Granta

Searle, J. (1999). *Mind, language, and society.* London: Phoenix.

Sennett, R. (1977). *The fall of public man*. Cambridge: Cambridge Uinverisity Press.

Smith, B. (1999). Truth and the visual field, in J. Petitit, F. J. Varela, B. Pachoud, & J.-M. Roy (Eds.), *Naturalizing phenomenology* (pp. 317-329). Stanford: Stanford University Press.

Smith, B., & Varzi, A. (1999). The niche. *Noûs, 33* (2), 198-222.

Sonesson, G. (1978). *Tecken och handling. Från språkhandlingen till handlingens språk.* Lund: Doxa (Dissertation).

Sonesson, G. (1979). A plea for integral linguistics. In *Papers from the 5th Scandinavian Conference of Linguistics, II* (pp. 151-166). Stockholm: Almqvist & Wiksell International.

Sonesson, G. (1989). *Pictorial concepts. Inquiries into the semiotic heritage and its relevance for the analysis of the visual world.* Lund: ARIS/Lund University Press.

Sonesson, G. (1992). The semiotic function and the genesis of pictorial meaning. In E. Tarasti (Ed.), *Center/Periphery in representations and institutions. Proceedings from the Conference of The International Semiotics Institute, Imatra, Finland, July 16-21, 1990* (pp. 211-156). Imatra: Acta Semiotica Fennica.

Sonesson, G. (1993). Pictorial semiotics, Gestalt psychology, and the ecology of perception. *Semiotica 99* (3/4), 319-399.

Sonesson, G. (1994). Prolegomena to the semiotic analysis of prehistoric visual displays. In *Semiotica, 100* (3/4), 267-312.

Sonesson, G. (1996). An essay concerning images. From rhetoric to semiotics by way of ecological physics. *Semiotica, 109* (1/2), 41-140.

Sonesson, G. (1997). Approaches to the Lifeworld core of visual rhetoric. *VISIO, 1* (3), 49-76.

Sonesson, G. (1998). On the notion of text in cultural semiotics. Σημειωτικ'η. Trudy po znakovym sistemam. *Sign System Studies, 26,* 83-114.

Sonesson, G. (2001) From Semiosis to Ecology. In A. Quinn (Ed.), Cultural cognition and space Cognition/Cognition culturelle et cognition spatiale [Special issue]. *VISIO, 6* (2-3), 85-110.

Sonesson, G. (2006a). *Current issues in pictorial semiotics. Lecture one: The quadrature of the hermeneutic circle.* First conference of a series published online at the Semiotics Institute Online. Retrieved October 30, 2009 from http://www.chass.utoronto.ca/epc/srb/cyber/Sonesson1.pdf

Sonesson, G. (2006b) The meaning of meaning in biology and cognitive science. A semiotic reconstruction. *Sign System Studies, 34* (1), 135-213.

Sonesson, G. (2007). From the meaning of embodiment to the embodiment of meaning. In T. Ziemke, J. Zlatev, & R. Frank (Eds.), *Body, language, and mind* (pp. 85-110). Berlin & New York: Mouton de Gruyter.

Sonesson, G. (2009). New considerations on the proper study of man — and, marginally, some other animals. *Cognitive Semiotics, 4* (Spring), 134–169.

Sonesson, G. (In press a) Semiosis and the Elusive Final Interpretant of Understanding. In *Semiotica*, 169, 2010.

Sonesson, G. (in press b). Semiosis beyond Signs. On a Two or Three Missing Links on the Way to Human Beings. To be published in the *Acts* from Missing Links Conference, Copenhagen, November 22nd-23rd 2007.

Sonesson, G. & Zlatev, J. (2008) Conclusions of the SEDSU project. In Sinha, C, Sonesson, G, & Zlatev, J. (Eds.), *Signing up to the human.* (In preparation)

Stjernfelt, F. (2007). *Diagrammatology. An investigation on the borderline of phenomenology, ontology, and semiotics.* Dordrecht: Springer.

Spiegelberg, H. (1956). Husserl's and Peirce's phenomenologies: coincidence or interaction. *Philosophy and Phenomenological Research, 17,* 164-185.

Suchman, L. A. (1987). *Plans and situated actions: The problem of human-machine communication.* Cambridge: Cambridge University Press.

Thompson, E. (2007). *Mind in life: biology, phenomenology, and the sciences of mind.* Cambridge, MA: Belknap (Harvard University Press)

Tomasello, M. (1999) *The cultural origins of human cognition.* Cambridge, MA: Harvard University Press.

Tomasello, M. (2008) *Origins of human communication.* Cambridge, MA: The MIT Press.

Varela, F., Thompson, E. & Rosch, E. (1991). *The embodied mind: cognitive science and human experience.* Cambridge: The MIT Press.

Vico, G. (2004). *Opere di Giambattista Vico. 8, La scienza nuova 1730.* Bologna: Il Mulino

Vygotsky, L. (1962) *Thought and language.* Cambridge, MA: The MIT Press.

Yates, F. (1964). *Giordano Bruno and the Hermetic tradition.* London: Routledge & K. Paul.

Yates, F. (1966). *The art of memory.* London: Routledge & K. Paul.

Zahavi, D. (2003). *Husserl's phenomenology.* Stanford, CA: Stanford University Press.

Zlatev, J. (2003). Meaning = Life (+ Culture). An outline of a unified biocultural theory of meaning. *Evolution of communication, 4* (2), 253-296.

Zlatev, J. (2005) What's in a schema? Bodily mimesis and the grounding of language. In B. Hampe (Ed.), *From perception to meaning: Image schemas in cognitive linguistics* (pp. 323–342). Berlin: Mouton de Gruyter.

Zlatev, J. (2007). Embodiment, language and mimesis. In T. Ziemke, J. Zlatev, & R. Frank (Eds.), *Body, language and mind: Vol. 1. Embodiment* (pp. 297-338). Berlin: Mouton de Gruyter.

Zlatev, J. (2008). The co-evolution of intersubjectivity and bodily mimesis. In J. Zlatev, T. Racine, C. Sinha, & E. Itkonen (Eds.), *The shared mind: Perspectives on intersubjectivity* (pp. 215-244). Amsterdam: John Benjamins Publishing Company.

Zlatev, J. (2009). The semiotic hierarchy: Life, consciousness, signs and language. *Cognitive Semiotics, 4* (Spring), 170–201

Craft, D. (2009). *The Golden Horde* (detail). Microphoto: Substituted Resorcinol Melt Crystals; 34 x 21 cm.

Cybernetics and Human Knowing. Vol. 16, nos. 3-4, pp. 149-174

Levels of Meaning, Embodiment, and Communication

Jordan Zlatev[1]

Departing from the theoretical framework of the semiotic hierarchy (Zlatev, 2009), realizing a form of cognitive semiotics based on "integrating methods and theories developed in the disciplines of cognitive science with methods and theories developed in semiotics and the humanities" (cf. www.cognitivesemiotics.com) the paper analyzes the concepts of embodiment and communication along an evolutionary progression of four levels: biological, phenomenological, significational (sign-based) and extended/normative. Examples of human and animal communication are provided in order to clarify these distinct levels. Further, the concept of bodily mimesis and the model of the *mimesis hierarchy* (Zlatev, 2008; Zlatev & Andrén, 2009) that is predicated upon it are offered as conceptual and empirical tools in order to help explain the transitions leading to the two highest meaning levels: to sign use proper (from pre-sign meanings); and from this to normative, and eventually body-independent, objectified sign systems such as those of writing and mathematics. The leitmotif of the paper is the risk of either exaggerating or down playing the bodily bases of meaning and communication.

1. Introduction

The ways in which the (human) body shapes meaning and thought, and the various roles it plays in communication, have received a great deal of attention during the last few decades. However, behind the slogan-like notion of the "embodied mind" lie a plethora of different concepts and theoretical frameworks, and the extent to which they are mutually compatible remains unclear (cf. Ziemke, Zlatev and Frank 2007; Krois et al. 2007).

In this article, focusing on the idea of the embodiment of meaning, I outline four different kinds of embodiment: biological, phenomenological, significational (sign-based) and extended, corresponding to the four major levels of meaning in a particular cognitive semiotic framework, The Semiotic Hierarchy (Zlatev 2009). Since the notion of "communication", with considerably older ancestry than that of embodiment, is even more ambiguous, I then attempt to provide a definition and propose a framework distinguishing between four levels of communication, corresponding to the levels of meaning and embodiment outlined earlier, illustrating these with examples from the comparative and developmental literature.

Some conclusions that can be drawn from this analysis are that a purely biological view of embodiment can only account for the lowest level of communication. By complementing the biological perspective of the body with a phenomenological one (Husserl 1989 [1952]; Merleau-Ponty 1962), focusing on "the lived body" (*Leib*), we

---

1. Lund University, SOL, Centre for Cognitive Semiotics (CCS). Copenhagen Business School, Language, Cognition and Mentality (LaCoMe) Email: Jordan.Zlatev@ling.lu.se

can accommodate crucial dimensions of communication such as agency and intention. Furthermore, a phenomenological semiotics can provide us with a notion of *sign* or representation that is not "Cartesian" or "solipsist" in any way, but rather grounded in the acts of the lived body, and furthermore on the roles of symbolic artifacts and external representations (Sinha 1988, 2009; Donald 1991; Sonesson 2007a, 2007b; 2009).

Level- and stage-based analyses, however, remain only descriptive if they cannot tell us anything about the *transitions* between the levels, in either phylogeny or ontogeny. In this respect, the concept of *(bodily) mimesis* (Donald 1991; Zlatev 2005, 2007) can be seen to play an explanatory role. In particular, I will summarize the evolutionary and developmental model of the Mimesis Hierarchy (Zlatev 2008a, 2008b) and suggest that it can help explain the emergence of signification, and possibly even the "highest" levels of meaning and communication going beyond mimesis into conventional/normative representations such as those of language, and (eventually) external(ized) representations.

Finally, I point to a potential danger eminent in focusing too much on the "highest" level of meaning (communication and embodiment): a forgetting, and potentially devaluing of the foundational role of the real, living and lived human body in "bringing forth" a human - and humane - world.

2. Embodiment

2.1. What is "embodiment"?
As often rehearsed these days (e.g. Ziemke et al. 2007; Wallace et al. 2007), the "classical" information-processing paradigm within the cognitive sciences operated with a disembodied notion of the "mind/brain", working similarly to a digital computer (e.g. Chomsky 1965; Fodor 1981; Jackendoff 1983; Pinker 1994). In reaction to this conception and its many problems, the term "embodiment" has been something of a rallying call for those looking, for good reasons, for an alternative. Rhetorically, this is often framed as a "radical" paradigm shift.

> Radical embodiment ... [is] radically altering the subject matter and theoretical framework of cognitive science. (Clark 1999: 22)

> We propose a radically different view. We will argue that conceptual knowledge is embodied, that is, it is mapped within the sensory-motor system. (Gallese and Lakoff 2005: 456)

> Embodied cognition offers a radical shift in explanations of the human mind – a Copernican revolution in cognitive science – you might say, which emphasizes the way cognition is shaped by the body and its sensorimotor interaction with the world. (Lindblom and Ziemke 2007: 129-130).

Unfortunately, however, despite initial optimism (cf. Varela, Thompson and Rosch 1991), a widely accepted, coherent interdisciplinary theoretical framework for the study of human meaning, communication and thinking predicated on the notion of embodiment has not yet come about (cf. Zlatev 2007). The main reason behind this is the vagueness, or rather *ambiguity*, of the term "embodiment" itself. Based on an analysis of the use of the term in the literature(s), the cognitive psychologist Wilson (2002), for example, writes of "six views of embodied cognition" and the cognitive scientist Ziemke (2003) of "six different notions of embodiment", with the two sets only partially overlapping. The cognitive linguist Rohrer (2007: 348) states: "By my latest count the term "embodiment" can be used in at least twelve different important senses with respect to our cognition." Again, the way Rohrer cuts the embodied cake has little to do with the divisions made by Wilson and Ziemke.

Such *non*-convergence of meta-analysis is hardly surprising, since for any consistent "taxonomy" of embodiment, one first needs to ask: *what* (X) is it that is claimed to be embodied in *what* (Y)? The reader of the literature on embodiment will find at least the following terms substituting for the variables in the schema:

X = mind, language, meaning, concepts, thinking, the self...
Y = the biological body, robot bodies, neural networks, (sensorimotor areas in) the brain, sensorimotor interactions, image schemas, the cognitive unconscious, the phenomenal body, artifacts, practices, signs...

Related to the problem of ambiguity is that of *overextension*: Are *all* aspects of the (human) mind "embodied", and if so - in the same way? Some strong statements have been made to this effect:

> Image schematic structure is the basis for our understanding of *all* aspects of our perception and motor activities. [...] Conceptual Metaphor Theory proposes that *all* abstract conceptualization works via conceptual metaphor, conceptual metonymy, and a few other principles of imaginative extension. (Johnson and Rohrer 2007: 33, 38, my emphasis)

Again, it is not surprising that such claims have been met with skepticism (Haser 2005; Zlatev 2007; Sinha 2009).

2.2 The embodiment of meaning

In my own research (Zlatev 1997, 2003, 2005, 2007, 2009), as well as others (Sinha 1988, 2009, in press; Sonesson 2007a, 2007b; Emmeche 2007) a first step to making the concept of embodiment manageable has been first to narrow down X: *meaning*. Given the philosophical vexations that surround this latter notion, this may at first seem to be a rather unproductive move. However, with the rapprochement of ideas from (bio)semiotics, theoretical biology, phenomenology and "enactive" cognitive science over recent years (Gallagher 2005; Thompson 2007; Brier 2008; Stjernfelt 2008; Zahavi and Gallagher 2008), there are grounds for optimism that a "unified bio-cultural theory of meaning" (Zlatev 2003) may indeed be possible. What follows in the remainder of this section is an outline of one such approach, applied to distinguishing levels of embodiment.

Attempting to synthesize work from cybernetics to linguistics, Zlatev (2003) defined meaning as "the relationship between an *organism* and its *environment*, determined by [...] *value*"[2] (ibid: 258). More recently, this idea has been generalized into the framework of the Semiotic Hierarchy (Zlatev 2009). In brief: meaning exists if and only if there is: (a) *subject* S, (b) a subject-internal *value system* V and (c) a *world* W in which the subject (as being-in-the-world) is embedded. A particular phenomenon within the world (p) will have a given meaning M for S, according to the formulation (which has an expository purpose and should not be taken as indicating that meaning can be "measured") given in (1).

(1) $M(p, S) = W(p) * V(p, S)$

In other words, the meaning of a given phenomenon, for a given subject, will be determined by the "type" of world in which the phenomenon appears *and* the value of the phenomenon for the subject. If either p falls "outside" W, or its value for S is nil, p will be meaningless for S. Depending on the nature of (a), (b), and (c), four levels of meaning can be defined, summarized in Table 1.

Table 1: Summary of the four levels of meaning of The Semiotic Hierarchy (from Zlatev 2009)

Level	Subject	World	Value System
1	*Organism*	*Umwelt*	*Biological*
2	*Minimal self*	*Natural Lebenswelt*	*Phenomenal*
3	*Enculturated self*	*Cultural Lebenswelt*	*Significational (Sign-based)*
4	*Linguistic self*	*Universe of discourse*	*Normative*

2.2.1 The biological body

As proposed originally by von Uexküll (1982 [1940]) the most basic kind of subject S is a biological *organism*, even of the simplest kind. Its world W is that of the *Umwelt* – that part of the larger "environment" which is picked out by a value system V, which is either innately or through learning geared for the survival and reproduction of the organism. Only organisms (living systems), and not artificially created machines, have a set of closely related properties: *autopoiesis* (Maturana and Varela 1980), *identity-*

---

2. Note that the broad notion of *value* employed in (Zlatev 2003, 2009) and in the present article does not necessarily imply either culture or normativity, which apply only to specific (higher) levels of the Semiotic Hierarchy. Alternatively, one could attempt to extend the notion of normativity all the way down to the single biological cell, or even lower to the dynamics of "far-from-equilibrium systems" (Bickhard 2004), but (for reasons that I hope will become apparent) this is not the approach taken here.

world polarity (Thompson 2007), and an *intrinsic value system* (Edelman 1992), serving their own interests, rather than optimizing some externally defined function. No artificial system has these properties, and hence the only kind of body able to give rise to meaning in a non-metaphorical sense is the living, biological body: meaning is co-extensional with life (cf. Zlatev 2003).

2.2.2 The lived body

However, the subject of biology, the organism, is not necessarily an experiencing subject. The living body is not identical to *the lived (phenomenal) body* (Husserl's *Leib*) (cf. Husserl 1989 [1952]; Merleau-Ponty 1962). The relationship between the organism-subject and the phenomenon (e.g. the "smell" of the animal picked up by the tick in the famous example of von Uexküll), is intrinsically meaningful for the tick, but this is not a sufficient reason to conclude that the tick has subjective experience (*pace* the claims of von Uexküll). At the same time, the proto-intentional relationship inherent in the organism-*Umwelt* polarity, i.e. the *biological directedness* of the organism-subject toward phenomena which it "experiences" (due to its intrinsic value system) as meaningful, even if non-phenomenally, is a plausible ground for the emergence of consciousness (as primitive sentience) in evolution (cf. Popper 1962, 1992; Zlatev 2003; 2009; Thompson 2007).

Thus, on the level of phenomenal value/meaning, there is not only a biologically meaningful *Umwelt*, but a phenomenal *Lebenswelt* in which the subject finds himself immersed. The subject S is here a "minimal self" (Gallagher 2005), with (at least) affective and perceptual consciousness, which is *intentional* (i.e. directed) towards whatever is perceived. The second sense of "intention", related to agency and volition is related to having a *body image*, unifying (at least) haptic, proprioceptive and visual experience of one's own body (Gallagher 2005), giving "higher animals" (i.e. at least mammals) and infants (at least from 9 months of age) a "sense of self" (Stern 2000 [1985]), being capable of acting purposefully on its surroundings.

2.2.3 The significational body

Non-human animals (without special "enculturation" in a human culture and special programmes of reinforcement) and pre-9 month old-infants are, however, not capable of using and interpreting *signs*, defined as follows[3]:

A sign is present if and only if E (expression) *signifies* C (content) or at least R (referent), for subject S, so that:

- The relation is **asymmetrical** (E → C/R, not C/R ← E)
- E and C/R are **differentiated**: E is qualitatively different from C/R for S

---

3. The definition is based on what was originally given in Zlatev (2003: 275), but has been elaborated on the basis of influence from Sonesson (1989). Sonesson (2009) elaborates the definition of the sign even further, but I find the present one sufficient for current purposes.

• E and C/R are **connected**: in perceiving or enacting E, S indirectly perceives (or conceives of) C/R

This can be illustrated clearly in the case of pictorial signs. Investigating their (possible) understanding by great apes, Persson (2008) distinguishes between (a) "surface mode", in which only the marks of lines and colour are perceived (*Bildding*, in Husserl's terminology), (b) "reality mode", in which the picture is confused with the object it represents (*Sujet*), e.g. a banana and (c) "pictorial mode", in which the *Bildding* is seen as an expression with a certain kind of content (*Bildobjekt*) which can, but need not represent a particular object (*Sujet*). Only in the case of (c) does the subject (in this case, the ape) *see the picture as a sign*. This is clearly a representational relation, mediated by the picture's content. The sign concept of a phenomenological semiotics (Sonesson 1989; 2007a) can be seen as a generalization of this and can involve other semiotic resources such as gestures, symbolic play, pantomime, theatre – and of course language. In all these cases what is directly perceived is "non-thematic" (not focused on by the consciousness of the subject), while what is indirectly perceived (or conceived) is thematic (cf. Sonesson 2006, 2007a).

This concept of the sign, in contrast to a plethora of others, deserves to be called "phenomenological" since it is the consciousness of the subject (S) that makes both the differentiation and the connection possible (Zlatev 2008c, in press). While it is logically possible for "the sign function" to emerge individually, outside of acts of communication, as envisaged by Piaget (1945), signs are typically learned socially, through imitation and communication. They become stable, and eventually conventional (i.e. mutually known) in a "symbolic" culture. Thus, the subject S of this level is an *enculturated subject*, and the world W is not only the directly perceived and acted upon natural *Lebenswelt*, but also a *culturally mediated Lebenswelt*,[4] not replacing, but augmenting the first (cf. Zahavi 2003).

As will be discussed in Section 4, the role of the biological and lived bodies of the previous two levels appears to be fundamental for (the emergence of) signification (sign use) in two ways: through (whole body) imitation (Piaget 1945; Donald 1991; 2001; Zlatev 2007), and the most basic forms of sign use in ontogeny and possibly in evolution: iconic and deictic gestures.

2.2.4. The extended body
At the highest level of meaning (and latest in evolution/history and ontogeny), the Lifeworld of subjects as "linguistic selves" is extended to include not only the pre-sign meanings (e.g. those of direct perception) and pre-linguistic signs (e.g. mimetic rituals), but all those denizens of Popper's (1962, 1992) "world 3": cultural beliefs, myths, scientific theories, political ideologies, novels, poems, internet forums, blogs

---

4. Note, however, that not all aspects of culture imply signification (sign use), cf. Sonesson (2009). Still, signs are the most salient part of human cultures, and thus, as an approximation, a cultural *Lebenswelt* can be said to correspond to a "significational" one.

etc. which are made possible by language. This can be called, following Sinha's (2004) use of the concept, "a universe of discourse". A key aspect of meaning on this level, absent (in any fully realized way) earlier is *normativity*: the meanings expressed by language and its many derivative forms are communicated in ways that obey public, commonly known criteria of correctness, or "rules" (Wittgenstein 1953; Itkonen 1978, 2003, 2008; Zlatev 2008a).

But in which way does meaning on this level relate to "embodiment"? Unlike the self-evident role of the body (biological, lived/imagined and significational/ expressive) on the previous three levels, it seems that with the ascent of language, and especially external representations such as notions, pictures and diagrams, the role of the human body here is relatively marginal. Thus, in one sense, one can argue that meaning at this level becomes "dis-embodied". But we could also describe this as a matter of "extended embodiment... aspects and features of the experientially or ecologically significant, non-corporeal world" (Sinha and Jensen de López 2000: 24). Normativity is a property not only of language, but of all cultural artifacts (from chairs to bank notes), with their canonical rules (Sinha 1988) and status functions (Searle 1995). Some of these meanings may be analyzed in terms of "cultural affordances" (Sonesson 2009), but others – most clearly (again) notations and diagrams – are (systems of) signs, where the sign vehicles have as Sonesson puts it (2007a, 2007b), metaphorically, "gained a body of their own":

> ... as Husserl (1962a: 365-386) recognized in his study of the origin of geometry, for the idealization to be complete, its products have to be "embodied" in some kind of notational system, because only in that way can they gain a stable, public existence in a domain that is completely separate from their instantiations in the practical situations of the Lifeworld. (Sonesson 2007a: 87f)

Thus, an exclusive focus on language and the "universe of discourse" for the highest level of the Semiotic Hierarchy might be misleading, and we could use the term "extended body" to stand for all those modes of meaning and communication that both transcend the limits of human embodiment, and link bodily experience to the wider world of culture, in a global "semiosphere" (Lotmann 1990).

3. Communication in the perspective of cognitive semiotics

3.1 What is "communication"?

Dance and Larson (1976, Appendix A) list 126 different definitions answering this question... In an influential attempt to "clarify this muddy concept by outlining a number of basic elements used to distinguish communication" (Littlejohn 1999: 6), Dance (1970) has singled out three dimensions according to which concepts of communication differ, as discussed by the comprehensive overview volume *Theories of Human Communication* (Littlejohn 1999), from which the definitions (2-7) are taken ibid: 6-7). The first dimension is *generality*, with (2) being (arguably) much too general, and (3) clearly much too concrete. The second dimension is "intentionality", or rather *purposefulness*, with (4) requiring "intent", and (5) not. The third concerns whether definitions *presume accuracy, or success*, of communication, or not. The definition in (6) does so (along with the over-restrictive requirement that interchange be "verbal"), while (7) focuses on the "transmission", but does not require that the message is successfully received or understood.

(2)   Communication is the process that links discontinuous parts of the living world to one another. (Ruesch 1956: 462)

(3)   The means of sending military messages, orders, etc. as by telephone, telegraph, radio, couriers. (The American College Dictionary 1964: 224).

(4)   Those situations in which a source transmits a message to a receiver with conscious intent to affect the latter's behaviors. (Miller 1966: 92)

(5)   It is the process that makes common to two or several what was the monopoly of one. (Gode 1959: 5)

(6)   Communication is the verbal interchange of a thought or idea. (Hoben 1954: 77)

(7)   Communication is the transmission of information. (Berelson and Steiner 1964: 254)

Littlejohn (1999: 7) concludes: "Debates on the definition of communication are perennial." and quotes Dance approvingly: "We are trying to make the concept of 'communication' do too much work for us" (Dance 1970: 210, cited from Littlejohn 1999: 9). In his metatheoretical overview he provides a useful (if often somewhat superficial) division of "theories of communication" into five major groups, or "paradigms" (a) *structural-functional theories* deriving from system theory, semiotics, linguistics and discourse studies, (b) *cognitive and behavioral theories* from the cognitive and biological sciences, (c) *interactionist theories* from ethnomethodology and related forms of social studies, (d) *interpretive theories* from hermeneutics and phenomenology which "celebrate subjectivism or the preeminence of individual experience" (ibid: 15) and (e) *critical theories*, often based on Marxism, which "focus on issues of inequality and oppression" (ibid: 15). Most importantly, Littlejohn points out the different theories' strengths and weaknesses, and rather than "taking sides" shows that these (usually warring) camps focus on complementary aspects of communication: (a) and (c) on the social-cultural dimension, with (a) zeroing in on structures, while (c) on processes. Both (b) and (d) focus on the individual, but from different perspectives: (b) from the third-person perspective of "objective" observation, while (d) from the first-person perspective, typical for the humanities. Finally, while (a-d) all focus on understanding (and possibly explaining) communication, (e) attempts further to use such knowledge in order to change communicative practices and structures for the better.[5] It is hard not to agree with Littlejohn's conclusion: "These paradigms are more than theory types. They also

---

5.   Cf. the famous last of Marx's *Theses on Feuerbach* (Marx 1985): "Philosophers have hitherto only interpreted the world in various ways; the point is to change it".

embody philosophical commitments and values and reflect the kind of work that different theorists believe is important" (ibid: 16).

Given this diversity, it would be more than naïve to propose a "unified theory of communication", especially in the pages allotted. Still it is tempting to apply the synthetic theory of meaning outlined in the previous section in order to distinguish *levels of communication*, corresponding to the levels of meaning and embodiment presented in the previous section. After all, the ambitions of the emerging school or "paradigm" of cognitive semiotics, involve precisely the combination of the social-cultural and the individual approaches, the scientific "third-person" and the experiential "first-person" perspectives, as expressed on the home site of the journal with the same name www.cognitivesemiotics.com):

> The first of its kind, *Cognitive Semiotics* is a multidisciplinary journal devoted to high quality research, integrating methods and theories developed in the disciplines of cognitive science with methods and theories developed in semiotics and the humanities, with the ultimate aim of providing new insights into the realm of human signification and its manifestation in cultural practices.

Furthermore, it is my conviction that this is "the kind of the work" that is most "important", i.e. a multidisciplinary (or even "transdisciplinary", Brier 2008) approach, going against ideological borders that leave us with a one-sided (or else incoherent) world-view concerning phenomena lying at the very core of what defines us as human beings: meaning and communication. Thus, I venture to propose a 127th definition of communication, which lies in an intermediary position with respect to Dance's (1970) three dimensions, and Littlejohn's five paradigms:

(8)   Communication is the transmission of meanings through different (primarily bodily) expressions between two or more subjects.

With respect to generality, it is clearly less abstract than (2) by requiring that the communicating entities be *subjects*, rather than "parts of the living world" such as neurons or hormones. At the same time it is clearly general, rather than domain-specific, such as (3). With respect to "intentionality" it is non-committed, allowing subdivision into intended and non-intended transmission of meanings – in two different ways to be explicated below. As for "success", it uses the (often criticized) notion of "transmission", as in (6), but unlike it, it does not focus solely on the "sender", but on both parties ("between"). Also unlike (6), it does not concern only verbal meaning. At the same time, it does not require in all cases that the sender's *meaning* (rather than "information") to be *identical* with that of the receiver, as in (7), thus allowing for individual interpretation, and collective negotiation.

The potentially most problematic term in the definition is that of "meanings", but given the general meaning theory outlined in Section 2, this is neither used in a general, vague sense, nor is specific to a particular level or type of meaning, e.g. "verbal". In fact, meaning as defined in The Semiotic Hierarchy is a wider concept than communication, since it also involves "phenomena" in the world, especially in the first two levels of *Umwelt* (the meaningful environment) and *Lebenswelt* (the world accessible to consciousness) that are not produced by another subject. For communication in the sense of (8), the phenomenon seen (or understood) as meaningful by one subject is always produced by another subject, without or with volition (the distinction between levels 1 and 2), without or with signification (level 2 vs. level 3), without or with normative value (level 3 vs. level 4). We can use the generic term *communicative signal* for expressions-meanings on all these levels for the sake of convenience.[6]

These levels of communication correspond to those of the Semiotic Hierarchy, and the levels of embodiment described in Section 2. Furthermore, cutting across these levels, divisions can be made, depending on the kind of "materiality" of the communicative signals and the perceptual modalities of their perception. Table 2 summarizes this taxonomy, with categories of communicative signals to be explained in what follows.

Table 2: Levels of communication, corresponding to the four levels of meaning of the Semiotic Hierarchy, and the levels of embodiment (cf. Section 2), with categories of communicative signals from the different communicative modalities, or "channels"

Level	Subject	Embodiment	Bodily- Visible/Haptic	Vocal- Audible	Material- Visible/Audible
1	*Organism*	**Biological**	**Bodily reactions**	**Cries**	**Traces**
2	*Minimal self*	**Phenomenological**	**Intention- movements, Attention getters**	**Directed calls**	**Marks**
3	*Enculturated self*	**Significational**	**Gesture, pantomime**	**"Vocal gestures"**	**Early picture comprehension**
4	*Linguistic self*	**Normative/ Extended**	**Signed language**	**Spoken language**	**Writing, external representations**

3.2.1 Bodily reactions, cries and traces
On the lowest level of the hierarchy, behaviors of a given organism (Subject 1) affect the behavior of Subject 2 (the "receiver"), and thus serve as communicative signals, but are produced by Subject 1 (the "sender") automatically, as part of processes of bodily regulation. Through evolutionary processes that are fairly well understood (cf. Hauser 1996), these behaviors were selected not only for their regulative function but also for the way they affect other animals, and thus became communicative without any (necessary) mediation of awareness.

---

6. Though note that the terminology in the field differs widely. Clark (1996) uses the term in a way that corresponds to what I would call a "communicative sign" (level 3): "I use the term signal for any action by which one person means something for another person." (ibid: 13), and, for example Tylén (2008), follows this usage. Sinha (2004) on the other hand explicitly contrasts "signals" and "symbols" (corresponding to the distinction of levels 1+2 vs. 3 in the present account).

In the "bodily-visible" modality, piloerection (hair-raising) in mammals may be taken as an example of a completely involuntary *bodily reaction*,[7] which was then probably selected in evolution due to the fact that it makes the animal appear larger in size and thus more dangerous to an intruder. Many animal cries, such as dogs' barking, have an analogous role in the vocal-audible modality. While dog owners may give "rich interpretations" to these signals, and through reinforcement dogs can learn to suppress or modulate them, it has recently been convincingly argued that barking is the result of the tension dogs feel when placed in a potentially threatening situation from which they cannot, or will not for conflicting motivations, flee – a reaction that played a key role in their domestication by our ancestors over the past 10,000 years (Lord et al. 2009). Finally, at least partially "extra-bodily" non-volitional communicative signals such as urination – in Table 2 called "traces" – by a variety of territorial animals, including certain fish (Almeida et al. 2005), serve as biological "status" signals to competitors and potential mates.

What is common to these three types of communicative signals is that they serve as "symptoms" of the biological state of S1, "honestly" (sexual status, "mobbing") or not, (size) and are in principle not different for S2 from other aspects of the *Umwelt*. However, there is a complication. With the possible exception of the "chemical communication" of fish in the third example, these communicative signals, while produced completely "unconsciously", i.e. non-voluntarily, involve (almost certainly) phenomenal experience, both in their expression (e.g. the feelings provoking barking) and their "contents", e.g. fear. Thus, they would seem to involve not only basic, *Umwelt*-level meaning, but also a basic, pre-cultural and pre-significational *Lebenswelt*. This is indicative of the difficulties of separating the levels in actual cases, even though the distinctions can be maintained analytically.

3.2.2 Intention movements, attention getters, directed calls and marks

The communicative signals of the next level have been most extensively studied among the non-human primates, and especially the great apes: animals for which the presence of conscious experience and purposive action can hardly be in doubt (cf. Beshkar 2008; Zlatev 2009). Furthermore, the signals discussed here can be shown to be produced with the purpose of influencing the behavior of conspecifics, thus amounting to the definition given in (4). However, while being both "intentional" (i.e. volitional) and "communicative" this does not amount to a strong notion of *intentional communication* (Grice 1989), which requires a higher-level intention: not only to influence the behavior of the receiver, but an intention that the receiver understands the sender's intended meaning, a form of *third-level mentality*: "I wish that you understand that I mean X in producing Y" (Zlatev 2008a). This amounts to an

---

7.   "Piloerection starts when a stimulus such as cold or fright causes a discharge from the (involuntary) nervous system that triggers contraction of the little *arrectores pilorum* muscles. Contraction of these muscles elevates the hair follicles above the rest of the skin so the hair seems to "stands on end."" (MedicineNet.com)

understanding of signification (E stands for C/R, cf. Section 2), which would bring us to the next level of the hierarchy.

While all species of great apes have been shown capable to master signification given special tutoring and human enculturation: chimpanzees and bonobos (Savage-Rumbaugh et al. 1998), gorillas (Patterson 1980) and orangutans (Miles 1990), their spontaneous communicative signals are not true signs and do not amount to "intentional communication" as defined above (Deacon 1997, Tomasello 2008), though that claim has been contested (e.g. Savage-Rumbough 1998).

In the bodily-visual modality, there has lately been considerable interest in "ape gestures" (Call and Tomasello 2007; Pika 2008), and considerable individual and intra-species group variation, implying learning, has been shown. Leavens and colleagues (e.g. Leavens et al. 2008) have documented the widespread presence of spontaneous "pointing" in captive apes of all species, mostly to human receivers, but also among themselves in some special conditions, e.g. Savage-Rumbough (1986). However, Tomasello (2008, see also Pika 2008) persuasively argues that such "gestures" are qualitatively different from those of children in their second year of life. To put it simply (in the terminology of this paper), while children's deictic and iconic gestures are signs (level 3, see below), those of the apes are not, but can be categorized as either *intention movements* (IMs), *intention getters* (AGs) or a combination of both. IMs arise from so-called ontogenetic ritualization: e.g. a pulling of the other's body in a desired direction becomes toned down to a gentle tug with time, since S2 has learned to respond adequately to the initial part of S1's action, allowing it to become "stylized". Apes also demonstratively understand (even if "non-mentalistically") that the other needs to attend to such IMs for them to be efficient communicative signals, and hence when S2 is facing another direction, S1 will usually produce AGs – either in the bodily-haptic modality (touching, patting) or in the vocal-audible one (calling) in order to gain S2's attention prior to producing IMs. Tomasello (2008) argues that ape pointing, which in non-enculturated individuals is always to desired objects and most often food, arises precisely in this way.[8]

Vocal *calls* as AGs have already been mentioned as an example of communicative signals on this level within the vocal-audible modality. Unlike IMs and non-vocal AGs, however, most ape calls do not seem to be learned ("socially transmitted"), but species-general, "innate" signals, and hence Tomasello (2008) argues that ape bodily-visible communicative signals, and not calls, were the likely stepping stone for the evolution of language: an argument for the "gesture-first" position within the prolonged debate with "speech-first" theorists (cf. Johansson 2005). This is plausible, and consistent with the Mimesis Hierarchy model (see Section 4), but ape (and

---

8. Leavens et al. (2008) object, and protest of double standards for children and apes, but since their theoretical and methodological commitments are to a "methodological behaviorism" (ibid: 191) denying the existence of "mythical" unobservable entities such as intentions (unless they can be publically observed), their view makes it impossible to distinguish the first three levels of communication, and in particular level 2 and level 3. Unsurprisingly, Leavens and colleagues struggle hard to question the existence of *any* cognitive and representational differences between apes and human beings – apart from those that arise due to (public) language (Leavens and Racine 2009).

dolphin) vocal signals are not to be easily dismissed. In the case of the most studied non-human species in primatology, chimpanzees, calls have been shown to be of two types: "broadcast" and "proximal". The first, such as the "food-cry" are high-pitched, not addressed to anyone in particular, apparently involuntary (Deacon 1997), and while their communicative function is often agonistic (pro-social) rather than (only) antagonistic, they seem to be closer to the cries of level 1. The second type of calls are low-pitched, seem to be directed to particular individuals, voluntary produced and intended to have a particular effect, e.g. consoling a distressed relative. It has also recently been shown that the two types lead to different brain-activation patterns: more localized to the right-hemisphere for the directed, proximal calls (Taglialatela et al. 2008).

Also recently, the first case of (seemingly) spontaneous learning of a species-atypical vocal sound by a captive orangutan has been reported, showing "...a new aspect of great ape vocal learning by providing data that an orangutan has spontaneously (without any training) acquired a human whistle and can modulate the duration and number of whistles to copy a human model" (Wich et al. 2008). Therefore, it seems that Tomasello (2008) underestimates the complexity of ape vocal abilities, by treating them basically as a level 1 phenomenon (cries): non-voluntary and unlearned.

However, what was stated in the beginning of this subsection, that both bodily and vocal spontaneous animal signals are not communicative *signs*, remains unchallenged. The calls signaling different types of predators (leopard, eagle, snake) produced by macaques, which received much attention at the beginning of the 1990s, are now nearly unanimously agreed to be "broadcast" signals, serving their communicative functions, without being either learned or intentional (both in the sense of "voluntary" and "directed") to another (cf. Cheney and Seyfarth 2005). Surprisingly, an interesting case for possible spontaneous sign use by non-humans, made by Savage-Rumbaugh (1998) has not received much attention in the literature. It concerns the third type of modality: the external-visible one. During troop migrations, wild bonobos have been observed to break and leave branches at path crossings, possibly signaling to other members of the troop following them the direction that they had taken. The suggestion is that given the ecological context (dense vegetation preventing visual contact, predators that would be informed in the case of vocal signaling), this was a strategy consciously chosen by some individual troop members, and then became socially transmitted. If this interpretation were to be confirmed with more cases and better documentation, the branches would be almost literally "pointers" (i.e. deictic signs) fulfilling the conditions for signification and intentional communication given earlier. Unsurprisingly, Tomasello (1999) is skeptical, since breaking tree branches and dragging them is part of the behavioral repertoire of the species, and is used for a variety of "display" functions. Still, this is a good example, if only as a "thought experiment" alerting us that communication "in the wild" could take forms and modalities that are not readily apparent to us. In Table 2, I take (again) an intermediary position between Tomasello and Savage-Rumbaugh, calling the branches along the path *marks*: it is at least possible that the natural branch breaking behavior may become "ritualized" in a particular troop, so that some members voluntarily leave them along the way so as to influence the behavior of those following in a (literally) desired direction, though without involving intentional communication proper, i.e. involving a higher-level intention for their intended meaning to be understood.

3.2.4 Language: signed, spoken and systemic external representations

Given the lack of clear evidence for spontaneous sign use in non-human animals, and the highly circumstantial, and much debated evidence from "bones and stones", for good examples of pre-linguistic, but nevertheless significational meaning and communication, we need to turn to early childhood. To define the border between "prelinguistic" and the "linguistic" child is, of course, not unproblematic. The first words appear around the first birthday, but developmental psychologists from Vygotsky (1962) to the present disagree on whether they are truly "symbolic" (i.e. signs as here defined) rather than "indices" (Piaget 1945), which could be understood as level 2 communicative signals, i.e. associative pairing between a vocalization and a desired object or event. It is first with the "vocabulary spurt" around the middle of the second year, and clearly by 20 months that it is generally uncontested that the child has made his entry into language.

Considering the period between 9 and 18 months, on the other hand, there is indisputable evidence that the (typically developing) child has become a sign user, foremost in the bodily-visible modality (Piaget 1945; Bates, Camaioni and Volterra 1975; Bates 1979; Acredolo and Goodwyn 1990; Carpenter et al. 1998; Liszkowski et al. 2004; Blake et al. 2003). As suggested below (Section 4.2), however, this period can be subdivided in two stages. The child's first bodily communicative signals are also "dyadic" (e.g. raising the hands to express the wish to be picked up) and when "triadic", function as requests for objects. Thus they resemble the "gestures" of the great apes discussed in the previous sub-section. In a recent review, Pika (2008) asks a pertinent question "Gestures of apes and pre-linguistic human children: Similar or different?" and concludes that there are both similarities and differences: "Many human gestures are … used to direct the attention and mental states of others to outside entities… Apes also gesture… but use these communicative means mainly as effective procedures in dyadic interactions to request action from others" (ibid: 131-132). While the ontogenetic progression needs to be more carefully studied, especially in a cross-cultural perspective, it seems that the gestures that are specific for human children and which "direct the attention of others to some third entity, simply for the sake of sharing interest in it or commenting on it" (ibid: 131) appear clearly from about 13-14 months. It is in part a terminological issue, but in Table 2, I reserve the term *gesture* for those (human-specific) bodily expressions which (a) "stand" for a specific meaning, an actual or imagined object, action or event, and (b) in which this sign relationship is intentionally communicated. It is possible to have (a) without (b), as in "private" symbolic play or reenactment, but (b) clearly requires (a). Since the relationship between expression and meaning is not (yet) conventional, there are two ways in which the meaning can be "transmitted": through resemblance (iconic gestures, pantomimes[9]) and declarative (as opposed to imperative) pointing.

What about the other two modalities on this level? While this is controversial, I would suggest that prior to the vocabulary spurt around the middle of the second year, the child's first "words" serve a subordinate role to gestural communication, as a supplement to the multimodal communicative signal (Clark 1996). Their conventional referential function is not yet clear to the child and they serve as *"vocal gestures"* – a role more sophisticated than that of directed calls (level 2), but not yet part of a linguistic system (level 4). This is consistant with the growing acceptance of the view of "gesture as the cradle of speech" (cf. Acredolo and Goodwyn 1990; Iverson and Goldin-Meadow 1998; Lock and Zukow-Goldring in press).

Finally, a similar development seems to occur in the modality of external-visual representations during the period 9-20 months. Children's motoric skills are not yet mature in order to be able to produce pictorial signs (i.e. representational drawings) even by the end of this period. But if we look at studies of picture perception and understanding, we find a transition from a "reality mode" (cf. Persson 2008) in which the picture is confused with the object depicted, to the beginning of a "pictorial mode" in which the picture is understood to be a pictorial sign (at least by 18 months), though difficulties in establishing (and maintaining in memory) such so called "dual representation" (cf. DeLoache 2004) persist until the end of the third year, depending on the type of sign vehicle used and on the nature of the experimental paradigm.

3.2.4 Language: signed, spoken and systemic external representations

Following a transitional period of one word utterances supplemented non-redundantly with gestures to form word-gesture combinations (Iverson and Goldin-Meadow 1998), toward the end of the second year most children start quickly becoming increasingly proficient in combining linguistic expressions, and learning their internal relationships. Depending on their social environment more than on their perceptual capacities, this takes place either in the spoken or in "bodily-visible" modality, i.e. they become at first apprentices and eventually masters of either a spoken or a signed language. In some respects this is a "constructive" process (Tomasello 2003), and studies of the spontaneous emergence of Nicaraguan Sign Language (NSL) among deaf school-children who were being taught Spanish through lip-reading and writing (e.g. Senghas, Kita and Özyürek 2004) have shown that children not only spontaneously acquire an existing language, but are capable of co-creating one across several generations ("cohorts") of interacting signers. Still, common to both the

---

9. There is no clear difference between (preverbal) iconic gestures and (communicative) pantomime: both are performed from a "character viewpoint" rather than an "observer viewpoint" (McNeill 2005) or a "first-person perspective" rather than a "third-person perspective" (Zlatev and Andrén 2009). In other still other terms, they are *enactive* rather than *depictive* (note that a picture is typically produced from an observer viewpoint). It is only with the emergence of language and especially with the "gesture explosion" around 4 years of age (McNeill 2005) that an analytic distinction between co-speech gestures and pantomime becomes necessary.

"acquisition" and the "construction" perspectives is that what is ultimately established is a "socially-shared symbolic system" (Nelson and Shaw 2002) or a "conventional-normative semiotic system for communication and thought" (Zlatev 2008a). Such definitions of "language" include the two key features distinguishing language from pre-verbal gestures, (from which it gradually emerged in the case of NSL), as well as most representational images: (a) *conventionality*, in the sense of signs and their relations being commonly known and normative (Itkonen 2003, 2008) and (b) *systematicity*, most evident (and studied) on the sentence level as "grammar", but also on the level of discourse.

With literacy, to which children are introduced from 3 to 7 years of age, depending on the educational practices of different (literate) societies, a "universe of discourse" opens up (cf. Section 2.2.4). "Externally embodied" signs, and sign complexes such as (verbal) texts, (complex) pictures, numerical and graphic representations – in various combinations and media – constitute a considerable, if not the major part of the meaningful world for an increasing number of people, in our increasingly technologized societies. At the same time, the "lower" levels of meaning and communication continue to operate in parallel, and it would be both an intellectual mistake, committed by representatives of post-structuralism such as Derrida (1976), and a grave social mistake to de-value them, by either playing down their importance, or else attempting to assimilate them to the higher levels.

4. The Mimesis Hierarchy and transitions between the levels

As stated in the introduction, conceptual analyses (with empirical support) of meaning, embodiment and communication such as those presented in the previous two sections may be important as general theoretical frameworks, but they remain fundamentally descriptive, unless complemented with some sort of explanation of the transitions between the different levels. In other words, we need to take either an explicit evolutionary perspective, and consider how qualitatively new forms of meaning and communication *could* have emerged on our planet, or a developmental perspective and study how such changes can take place in children during the first years of life.

Perhaps most difficult is to address the first transition: from biology to experience (sentience, intentionality), but by framing the problem in a way similar to the approach here, i.e. as a "body-body problem" (i.e. the relation between biological body and the phenomenal body), rather than as the classical "mind-body problem", Thompson (2007) has shown that it is both conceptually and empirically tractable (cf. also Zlatev 2009). It is to the next two transitions that I now turn: from experience to signification (sign use), and from there to language, using the concept of bodily mimesis (Zlatev 2005; 2007; 2008a), looking first at evolution and then at development.

4.1. The mimetic origins of signification and language

How did the human capacity for signification (sign use) arise? Apart from differences in language capacity and other forms of sign use, many studies in comparative psychology have established important differences in *imitative capacity* and *intersubjectivity*, i.e. the ability to share and eventually to understand the experiences of others, between us and our closest relatives, the great apes (Donald 1991; Tomasello 1999, 2008; Zlatev et al. 2006; Persson 2008). Table 3 summarizes some of these. As can be seen, chimpanzees (and bonobos when tested) perform more or less similarly to children, except in imitation of action and mutual gaze (between mothers and infants), while gorillas and orang-utans differ in every respect (when tested) except for the recognition of being imitated. This last capacity has very recently been shown even for capuchin monkeys (Paukner et al. 2009).

Table 3: Comparative results concerning sign use (picture comprehension), intersubjectivity (food sharing, contagion, and mutual gaze) and imitation (on actions, and recognition of being imitated) from the SEDSU project: + = positive results, - = negative results, (+) = mixed results, ? = results lacking.

	Children (2 years)	Chimpanzees	Bonobos	Gorillas	Orangutans
Picture comprehension	+	+	+	-	?
Imitating actions	+	(+)	?	-	-
Food sharing	+	+	+	-	-
Yawning contagion	?	+	?	-	-
Mutual gaze	+	(+)	(+)	-	?
Recognizing being imitated	+	+	+	+	+

One interpretation consistent with such findings is that the trend for selection of social-cognitive skills in the hominid line gave rise (in early *Homo*) to capacities for intersubjectivity and imitation that are unique in the animal world. These essentially co-evolved and eventually gave rise to bodily signification (Zlatev 2008a, 2008b). Donald (1991) first proposed the general notion of *mimesis* as a "missing link" between ape-like "episodic" cognition and culture, and human "mythic" culture. I have endeavoured to elaborate this into a model, The Mimesis Hierarchy, consisting of 5 levels (Table 4), which can be seen to follow naturally upon each other, as suggested below.

Table 4: The five levels of The Mimesis Hierarchy (cf. Zlatev 2008a, 2008b)

Level	Characterized by Acts Which Are
Proto-mimesis	based on a cross-modal mapping between exteroception (normally dominated by vision) and proprioception (normally dominated by kinesthetics)
Dyadic mimesis	under conscious control and corresponding—either iconically or indexically—to some action, object or event, and at the same time being differentiated from it
Triadic mimesis	intended to stand for some action, object or event for an addressee (and for the addressee to recognize this intention)
Post-mimesis 1 (Protolanguage)	conventional-normative
Post-mimesis 2 (Language)	divided (semi)compositionally into meaningful sub-acts that systematically relate to other similar acts (as in grammar)

Proto-mimesis, implicated in acts of mutual attention and the awareness of others' feelings, is shared with many non-human primates, but was further selected for in our "ultra-social" species. This gave rise to distinctively human skills of *dyadic mimesis*, the ability to map between one's own body and that of others in a more detached and differentiated way. This allowed understanding others' emotions, shared attention and (non-communicative) intentions through a (conscious) process of "projection": what would I see/feel/wish if I were you. Still, as shown in Table 3, at least chimpanzees (and bobobos) have some spontaneous capabilities in this respect.

The more qualitative step in human cognitive-semiotic evolution involves *triadic mimesis*, implying having and understanding others' communicative intentions. This requires third-order mentality: "I want you to do X (e.g. share attention on an object) by recognizing my intention that you do this" from the sender's perspective and "I understand that you want me to do X" from the recipient's. Triadic mimesis is clearly difficult for apes to attain, especially in natural conditions. However, through enculturation and especially through extensive sign use, some understanding of communicative intentions seems to be within the reach of apes' "Zone of Proximal Development" (Donald 2001), and thus possibly in the common human-chimpanzee ancestor.

Post-mimesis 1, or protolanguage, which implies some understanding of semantic conventions, could have emerged as typified gestures. Post-mimesis 2, which is identical to language, has on top of everything else the command of a conventional/normative *system* for communication and *thought*. With this level the ultimate payoff of using the same system for communication and thought arises, giving us the cognitive benefits of (logical) reasoning, inference, long-term planning etc. that we take pride in as a species.

Thus, bodily mimesis – in its proto, dyadic and triadic forms – could possibly be a (and possibly *the*) major factor in the evolution of human intersubjectivity and communication. Signification would have been both a result and a driving force in the development of an understanding of third-order mentality, and eventually language. The topic of the evolution of language is currently intensely investigated and hotly debated (cf. Deacon 1997; Christiansen and Kirby 2003; Johansson 2005; Burling 2005; Tomasello 2008). The Mimesis Hierarchy model is in line with "gestural origins" theories (e.g. Donald 1991; Corballis 2002; Arbib 2005; Tomasello 2008). The transition from the manual-brachial to the vocal modality, often presented as a problem for such theories, could have occurred gradually over 1.5 million years from *H. ergaster* to *H. sapiens* as "vocal gestures" became increasingly recruited to supplement, rather than to replace gestural communication. By the appearance of *H. sapiens* in Africa about 200,000 years ago, this process would have been firmly established, and with this at least the emergence of an integrated gestural-vocal protolanguage (McNeill 2005; Zlatev 2008b). Through processes of cultural, rather than biological, evolution this gave rise to the multitude of languages we know today. Note that this scenario does not assume that *H. erectu*s had "gestural language", since triadic mimesis (pointing and pantomime) lacks the normative aspect inherent in language. Thus, the proposal is less vulnerable to the objection that evolutionary drift would have led to languages with predominantly manual-brachial signs, as in modern signed languages.

4.2 The Mimesis Hierarchy in children's cognitive-semiotic development
The Mimesis Hierarchy can also be interpreted as a stage-model of ontogenetic development (Zlatev and Andrén 2009), as summarized in Table 5. Each of the five successive stages can be defined through the *clear* attainment of a previously unavailable (cognitive) semiotic capacity. Unlike classical stage models in the spirit of Piaget where each consecutive stage brings with it total reorganization[10], this is a "layered model" (Stern 2000 [1985]) where earlier capacities continue to co-exist with newer ones, which subsume but do not abolish their predecessors. Each of the novel capacities defining the different stages typically make their entry somewhat earlier than specified in Table 5, but it takes time before they generalize beyond the first "islands" of use.

---

10. This received interpretation of Piaget's developmental model is somewhat simplified, since Piaget recognized that figurative, "preoperational" thinking subsists even in the later "operational" periods.

Table 5: The stages of the Mimesis Hierarchy, applied to child development

	Label	Novel capacity	Cognitive/communicative skills	Approx. age
1	Proto-mimesis	Mapping between exteroception and proprioception	- emotional and attentional contagion - neonatal imitation - mutual gaze	0-8 m
2	Dyadic mimesis	volition and representation	- imitation - imperative pointing - shared attention	9-13 m
3	Triadic mimesis	communicative signs	- declarative pointing - iconic gestures - (full) joint attention	14-19 m
4	Protolanguage	conventionality/ normativity	- one-word utterances - holophrases	20-27 m
5	Language	semiotic systematicity	- spoken or signed language	28 m-

Stage 1 (*proto-mimesis*) rests on a special form of active perception in which (dynamic) aspects of the environment – especially the actions of con-specifics – are mapped onto one's own bodily actions and sensations. Such "self-other matching" (Barresi and Moore 2008) is well-testified in infants. It makes possible the experience of "human-scale" meaningful physical and social aspects of the Lifeworld, e.g. to distinguish between inanimate objects, animals and persons, and to communicate (above all) affective states, via neonatal imitation (Meltzoff and Moore 1977) and proto-conversations (Trevarthen 1979). Yet, until approximately 8 months of age, this is done without a clear differentiation between self and other, or a sense of agency.

Stage 2 (*dyadic mimesis*) occurs once a "sense of a core self" (Stern 2000 [1985]) in which the body is felt to be "one's own", under *volitional control*, and the self clearly different from others. This seems to take place around the age of 9 months.[11] This makes the lack of direct control of "others' bodies" apparent, and along with that the need to communicate something that is *not* shared to others. This is initially done through communicative signals such as intention movements and attention getters (cf. Section 3.2.2). But increased bodily control, combined with differentiation from others also allows a surge in imitation (of novel actions and events) and with time to use the body as a true *representation* or *sign* of something else, i.e. signification. Piaget (1945) offers the example of an infant opening and closing her mouth to model the opening and closing of a matchbox, which would be an example of an *iconic* correspondence (i.e. based on similarity) between the act and the object of attention. Children's acts of pointing for themselves in order to help guide their attention (Bates, Camaioni and Volterra 1975), emerging around 10-11 months would qualify as *indexical* (and more specifically *deictic*) mimetic acts. But note that neither of these

---

11. While researchers disagree on its nature, there is considerable agreement for a (qualitative) transition in development around 8-9 months (e.g. Trevarthen and Hubley 1978; Kaye 1982; Tomasello 1999).

examples is communicative. Children at this stage do begin to point also "imperatively" for others, but even though this is literally "triadic" (since it involves three entities), it does not imply that children are using these gestures as communicative *signs*, expecting that the addressee will understand their (shared) meaning.

Stage 3 (*triadic mimesis*), which (as with the other stages) may begin somewhat earlier, becomes established around 14 months (Bates 1979; Liszkowski et al. 2004; Blake et al. 2003) and introduces precisely this: the intersubjective (self-other matching and differentiation) and representational (expression-content correspondences) abilities developed earlier are merged and become true communicative signs. The three-part relationship between (i) self-initiated mimetic gesture, (ii) its meaning and (iii) the receiver of the intended meaning is what justifies calling this "triadic mimesis". An example of an iconic mimetic sign is the *miming* of an action in addressee-directed "symbolic play". *Declarative pointing*, which is qualitatively distinct from imperative pointing (cf. Section 3.2.3), combines deixis and iconicity, since the motion and direction of the hand resembles the intended direction of attention of the addressee. Signs at this stage may have the same meaning for child and addressee (if communication succeeds), but they are *not known* to have the same meaning.

Stage 4 (*protolanguage*), from approximately 20 to 27 months, brings along a more or less *explicit understanding* (insight) that the meaning of the sign (gesture or word) is common to oneself and the addressee, i.e. the sign's *conventionality*. This is closely related with understanding that there is a "correct" way to express something, i.e. the dawn of a conception of *normativity*. With this, the iconic and/or indexical motivation – or "ground" (Sonesson 2007a) – of the sign loses much of its function, allowing the relationship between expression and content to become increasingly *arbitrary*.[12] The child at this stage engages in a gestural-verbal protolanguage, which still largely lacks grammatical organization. The transition to this stage is marked most clearly by the "vocabulary spurt", which usually starts earlier, but is clear by 20 months.[13]

Finally, Stage 5 (*language*) introduces *semiotic systematicity*, involving hierarchical relations between composite and simple signs (corresponding to what is usually referred to as "compositionality"), and furthermore relations to other signs. This corresponds to the basic mastery of a public language (spoken or signed). Children make this transition at different ages, but 27 months is an approximate average.

Zlatev and Andrén (2009) investigated the development of so-called acts of bodily communication (ABCs) in three Swedish and three Thai children, between 18 and 27 months of age, using a transcribed and video-linked corpus of spontaneous adult-

---

12. Therefore arbitrariness and conventionality are related, but not synonymous notions. In a conventional language such as ASL, up to 50% of signs are more or less iconic (Woll and Kyle 2004).

13. "At first their rate of vocabulary is very slow, but one typically sees a "burst" or acceleration in the rate of vocabulary growth somewhere between 16-20 months" (Bates 2003: 15).

infant interactions. Approximately 1600 such acts were identified over the period, and analyzed using a semiotics-based coding system, distinguishing (on the highest level, with sub-categories not discussed here) between *deictic* (DEI), *iconic* (ICO) and *emblematic* (i.e. conventional) (EMB) "components", since one and the same ABC need not include only one type of semiotic ground. One of the most interesting findings of this study was the following:

> When viewing the children from both cultures as a single group, some general developmental patterns appeared. In particular, there was *evidence for a transition around 20 months*, when DEI components (in association with deictic expressions and nominals) peaked, along with a dip in ICO components, and a rather sudden increase of EMB components (consisting to a considerable degree in *nod-yes* and *headshake-no*). (Zlatev and Andrén 2009: 396).

Consistent with the Mimesis Hierarchy model, this is evidence for a transition from triadic mimesis to protolanguage (cf. Table 5 above). Since changes in the measures correlated in time (albeit for the group as a whole), and the transition seemed to be relatively "discrete", the findings can be interpreted as evidence for an *insight* of semiotic normativity, occurring cross-culturally around 20 months. In other words, that children at around this age develop the meta-cognitive awareness that an E(xpression) can not only be used to mean a certain C(ontent), but that it is *appropriately* used in their (mini)culture to do so. The fact that the "vocabulary spurt" typically begins around this time may be (at least in part) a result of this "symbolic insight".

5. Final remarks

At the onset of this decade and century, assessing the status quo within linguistic and philosophical semantics, cognitive science, and my (at that time somewhat sketchy) knowledge of semiotics, I made the following pessimistic pronouncement:

> Our conception of *meaning* has become increasingly fragmented, along with much else in the increasing 'postmodernization' of our worldview. The trenches run deep between different kinds of meaning theories: mentalist, behaviorist, (neural) reductionist, (social) constructivist, functionalist, formalist, computationalist, deflationist… And they are so deep that a rational debate between the different camps seems impossible. The concept is treated not only differently but *incommensurably* within the different disciplines (Zlatev 2003: 253).

This served as the motivation for attempting to formulate "an outline of a unified bio-cultural theory of meaning", giving a foundational place to life (rather than machines), and proposing "hierarchies" of meaning in evolution and development, which in a broadly continuous framework could also accommodate qualitative changes. At that time, my ambitions seemed somewhat premature, but since then, several impressive attempts at providing integrational theories of meaning have been proposed (Emmeche 2007; Sonesson 2007a; Stjernfelt 2007, Brier 2008) as well as a rapprochement between phenomenology and cognitive science (Thomspon 2007;

Gallagher and Zahavi 2008; Schmicking and Gallagher in press). The appearance of the journal *Cognitive Semiotics* on the scene can be seen as a reflection of the same need to counter the fragmentation described in the quotation above. The Semiotic Hierarchy (Zlatev 2009) and its extension to the notion of "embodiment" and communication explored in this chapter are in line with these developments. There are both similarities and differences between my proposals and those of colleagues, but these are to be explored in some other context.

I wish to conclude by emphasizing what seems common to all those mentioned in the previous paragraph – an effort to assert "the primacy of the body", but without falling into any form of biological reductionism in which the body (with focus on the brain) is treated as kind of physical object, a sophisticated machine. Another common motivation in theoretical frameworks such as the present one, as well as "embodied dynamism" (Thompson 2007) and "cybersemiotics" (Brier 2008) is a desire to point out that "higher levels" of meaning, communication and intersubjectivity presuppose lower ones: evolutionarily, developmentally, but also "synchronically". Meaning and communication are *rooted* in the biological, lived and significational bodies interacting with their respective "worlds" (cf. Table 1). This is important since neglecting it in theorizing leads to distorted accounts involving at one extreme beliefs in innate "language organs", and at another extreme, claims that "everything is a text".

Still much worse would be to neglect and devalue the body *in practice*: a cultural devaluing of the living and lived body in an over-technological society and "globalized" world. This could potentially lead to the experience of a vacuum of meaning, breakdown of communication, and ultimately self-destruction, on the individual or societal levels. The recent film *Babel* (2006) by the director Alejandro González Iñárritu portrays exactly this: how we are separated and antagonized by differences of language, custom, arbitrary norms, how a particular "cultural artefact" (a rifle) given as a token of gratitude turns into, by chance, the cause of mutual misery, how physical our suffering is as the result of torture, thirst, bullet wounds, suppressed sexuality... – but also how united we are as human beings in this suffering, and how our bodies are ultimately the only means of reaching out, and caring.

Acknowledgements

I wish to thank Mats Andrén, Göran Sonesson and an anonymous reviewer for comments on a previous version of this article.

References

Acredolo, L. and Goodwyn, S. (1990). Sign language among hearing infants: The spontaneous development of symbolic gestures. In V. Volterra and C.J. Erting (Eds.), *From Gesture to Language in Hearing and Deaf Children,* 68-78. Berlin: Springer.

Almeida, O.G., Miranda, A., Frade, P., Hubbard, P.C., Barata, E.N. and Canário, A.V.M. (2005) Urine as a Social Signal in the Mozambique Tilapia (*Oreochromis mossambicus*), *Chemical Senses* 30 (suppl 1): i309–i310, 2005

Barresi, J. and Moore, C. (2008). The neuroscience of social understanding. In J. Zlatev, T. Racine, C. Sinha, and E. Itkonen, *The Shared Mind: Perspectives on Intersubjectivity,* 39-66. Amsterdam: Benjamins.

Bates, E. (1979) *The Emergence of Symbols. Cognition and Communication in Infancy.* New York: Academic Press.

Bates, E. (2003) On the nature and nurture of language. UCSD, Unpublished report.

Bates, E., Camaioni, L., and Volterra, V. (1975). Performatives prior to speech. *Merrill-Palmer Quarterly* 21: 205-226.

Berelson, B., & Steiner, G. (1964). *Human Behavior.* New York: Harcourt, Brace and World.

Beshkar, M. (2008). Animal consciousness. *Journal of Consciousness Studies , 15* (3), pp. 5-34.

Bickhard, M. H. (2004). Process and emergence: Normative function and representation. *Axiomathes - An International Journal in Ontology and Cognitive Systems* 14:135-169.

Blake, J., Osborne, P., Cabral; M. and Gluck, P. 2003. The development of communicative gestures in Japanese infants. *First Language* 23(1): 3-20.

Brier, S. (2008). *Cybersemiotics: Why Information is not Enough.* Toronto: University of Toronto Press.

Burling, R. (2005). *The Talking Ape.* Oxford: Oxford University Press.

Call, J., & Tomasello, M. (2007). *The Gestural Communication of Apes and Monkeys.* Mahwah, NJ: Lawrence Erlbaum.

Carpenter, M., Nagell, K., & Tomasello, M. (1998). *Social cognition, joint attention, and communicative competence from 9 to 15 months.* Monographs of the Society of Research in Child Development 63(4).

Cheney, D. L., & Seyfarth, R. M. (2005). Constraints and preadaptations in the earliest stages of language evolution. *Linguistic Review , 22*, pp. 135-159.

Clark, A. (1999.). An embodied cognitive science? *Trends in Cognitive Sciences , 3* (9), pp. 345-351.

Chomsky, N. (1965). *Aspects of a Theory of Syntax.* Cambridge, Mass.: MIT Press.

Christiansen, M. & Kirby, S. (2003). *Language Evolution.* Oxford: Oxford University Press.

Clark, H. H. (1996). *Using Language.* Cambridge: Cambridge University Press.

Corballis, M. C. (2002). *From Hand to Mouth: The Origins of Language.* Princeton, NJ: Princeton University Press.

Dance, F. E. (1970). The "concept" of communication. *Journal of Communication , 20*, pp. 201-220.

Dance, F. E., & Larson, C. E. (1976). *The Foundations of Human Communication: A Theoretical Approach.* New York: Holt, Rinehart & Winston.

Deacon, T. (1997). *The Symbolic Species: The Co-evolution of Language and the Brain.* New York: Norton.

DeLoache, J.S. (2004). Becoming symbol-minded. *Trends in Cognitive Sciences* 8: 66–70.

Derrida, J. (1976). *Of Grammatology.* Baltimore: John Hopkins University Press.

Donald, M. (1991). *Origins of the Modern Mind: Three Stages in the Evolution of Culture and Cognition.* Cambridge, Mass.: Harvard University Press.

Donald, M. (2001). *A Mind so Rare: The Evolution of Human Consciousness.* New York: Norton.

Edelman, G. (1992). *Bright Air, Brilliant Fire: On the Matter of the Mind.* New York: Basic Books.

Emmeche, C. (2007). On the biosemiotics of embodiment and our human cyborg nature. In T. Ziemke, J. Zlatev, & R. Frank, *Body, Language, Mind: Volume 1, Embodiment* (pp. 411-430). Berlin: Mouton de Greyter.

Fodor, J. A. (1981). *Representations.* Cambridge, Mass.: MIT Press.

Gallese, V., & Lakoff, G. (2005). The brain's concepts: The role of the sensori-motor system in conceptual knowledge. *Cognitive Neuropsychology , 22*, pp. 445-479.

Gallagher, S. (2005). *How the Body Shapes the Mind.* Oxford: Oxford University Press.

Gallagher, S., & Zahavi, D. (2008). *The Phenomenological Mind: An Introduction to Philosophy of Mind and Cognitive Science.* London: Routledge.

Gode, A. (1959). What is communication. *Journal of Communication , 9*, pp. 3-20.

Grice, P. (1989). Meaning. In *Studies on the Way of Words* (pp. 213-223). Harvard: Harvard University Press.

Haser, V. (2005). *Metaphor, Metonymy and Experientialist Philosophy.* Berlin: Mouton de Gruyter.

Hauser, M. D. (1996). *The Evolution of Communication.* Cambridge, Mass: MIT Press.

Hoben, J. B. (1954). English communication at Colagte Re-examined. *Journal of Communication , 4*, pp. 72-92.

Husserl, E. (1989 [1952]). *Ideas Pertaining to a Pure Phenomenology and to a Phenomenological Philosophy, Second Book.* Dordrecht: Klewer.

Itkonen, E. (1978). *Grammatical Theory and Metascience: A Critical Inquiry into the Philosophical and Methodological Foundations of "Autonomous" Linguistics .* Amsterdam: Benjamins.

Itkonen, E. (2003). *What is Language? A Study in the Philosophy of Linguistics.* Turku: Turku University Press.

Itkonen, E. (2008). The central role of normativity for language and linguistics. In J. Zlatev, T. Racine, C. Sinha, & E. Itkonen, *The Shared Mind: Perspectives on Intersubjectivity,* 279-306. Amsterdam/Philadelphia: Benjamins.

Iverson, J.M. and Goldin-Meadow, S. (1998). *The Nature and Functions of Gesture in Children's Communication.* San Fransisco: Jossey-Bass Publishers.

Jackendoff, R. (1983). *Semantics and Cognition.* Cambridge, Mass.: The MIT Press.

Johnson, M., & Rohrer, T. (2007). We are live creatures: Embodiment, American Pragmatism and the cognitive organism. In T. Ziemke, J. Zlatev, & R. Frank, *Body, Language and Mind. Vol 1: Embodiment* (pp. 17-54). Berlin: Mouton de Gruyter.

Johansson, S. (2005). *Origins of Language: Constraints on Hypotheses.* Amsterdam: Benjamins.

Kaye, K. (1982). *The Mental and Social Life of Babies: How Parents Create Persons.* Chicago: University of Chicago Press.

Krois, J. M., Rosengre, M., Steidele, A., & Westerkamp, D. (2007). (Eds.) *Embodiment in Cognition and Culture.* Amsterdam: Benjamins.

Leavens, D. A., & Racine, T. (2009). Joint attention in apes and humans: Are humans unique? *Journal of Consciousness Studies , 16* (6-8), pp. 240-267.

Leavens, D. A., Hopkins, W. D., & Bard, K. A. (2008). The heterochronic origins of explicit reference. In J. Zlatev, T. Racine, C. Sinha, & I. Itkonen, *The Shared Mind: Perspectives on Intersubjectivity* (pp. 187-214). Amsterdam: Benjamins.

Lindblom, J., & Ziemke, T. (2007). Embodiment and social interaction: A cognitive science perspective. In T. Ziemke, J. Zlatev, & F. R, *Body, Language and Mind. Vol 1: Embodiment* (pp. 129-163). Berlin: Mouton de Gruyter.

Liszkowski, U., Carpenter, M., Henning, A., Striano, T., and Tomasello, M. (2004). Twelve-month-olds point to share attention and interest. *Developmental Science* 7: 297–307.

Littlejohn, S. E. (1999). *Theories of Human Communication.* Belmont, CA: Wadsworth Publishing Company.

Lock, A and Zukow-Goldromh. P. (in press) Prelinguistic communication. In G. J. Bremmer and T. D. Wachs (Eds) *Handbook of Infant Development, Second Edition.* Oxford: Blackwell.

Lotman, Y. (1990). *Universe of Mind. A Semiotic Theory of Culture.* New York: I.B. Tauris and Co.

Lord, K., Feinstein, M. and Coppinger, R. (2009) Barking and mobbing. *Behavioural Processes* 81 (3): 358-368.

McNeill, D. (2005). *Gesture and Thought.* Chicago: University of Chicago Press.

Maturana, H., & Varela, F. (1980). *Autopoiesis and Cognition - The Realization of the Living.* Dordrecht: Reidel.

Meltzoff, A.N. and Moore, M.K. (1977). Imitation of facial and manual gestures by human neonates. *Science* 198: 75-78.

Merleau-Ponty, M. (1962 [1945]). *Phenomenology of Perception.* London: Routledge.

Miles, H. L. (1990). The cognitive foundations for reference in a signing orangutan. In S. T. Parker, & K. R. Gibdon, *"Language" and Intelligence in Monkeys and Apes* (pp. 511-539). Cambridge: Cambridge University Press.

Miller, G. R. (1966). On defining communication: Another look. *Journal of communication , 16*, pp. 90-112.

Nelson, K. and Shaw, L. K. (2002). Developing a socially shared symbolic system. In *Language, Literacy and Cognitive Development,* J. Byrnes and E. Amseli (eds.), 27-57. Hillsdale, NJ: Lawrence Erlbaum

Patterson, F. (1980). Innovative use of language in a gorilla: A case study. In K. Nelson, *Children's Language. Vol. 2* (pp. 497-561). New York: Garnder Press.

Paukner, A. Suomi, S.J. Visalberghi, E., Ferrari, P.F. (2009) Capuchin monkeys display affiliation toward humans who imitate them, *Science* 325 (5942) pp. 880 – 883.

Persson, T. (2008). *Pictorial Primates: A Search for Iconic Abilities in Great Apes.* Lund: Lund University Cognitive Studies, 136.

Piaget, J. (1945). *La formation du symbole chez l'enfant.* Delachaux et Niestlé.

Pika, S. (2008). Gestures of apes and pre-linguistic human children: Similar or different? *First Language* 28 (2): 116-140.

Pinker, S. (1994). *The Language Instinct.* New York: William Morrow.

Popper, K. (1962). *Objective Knowledge.* Oxford: Oxford University Press.

Popper, K. (1992). *In Search of a Better World: Lectures and Essays from Thirty Years.* London: Routledge.

Ruesch, J. (1957). Technology and social communication. In L. Thayer, *Communication Theory and Research.* Springfield, IL: Thomas.

Savage-Rumbaugh, S. (1998). Scientific schizophrenia with regard to the language act. In J. Langer, & M. Killen, *Piaget, Evolution and Development* (pp. 145-169). Mahwah, NJ: Erlbaum.

Savage-Rumbaugh, S. (1998). Scientific schizophrenia with regard to the language act. In J. Langer, & M. Killen, *Piaget, Evolution and Development* (pp. 145-169). Mahwah, NJ: Erlbaum.

Schmicking, D. and Gallagher, S. (in press) (Eds.), *Handbook of Phenomenology and Cognitive Sciences.* Dordrecht and New York: Springer, 2009.

Searle, J. (1995). *The Construction of Social Reality.* London: Allen Lane.

Senghas, A., Kita, S. and Özyürek, A. (2004). Children creating core properties of language: Evidence from an emerging sign language in Nicaragua. *Science* 305: 1779-1782.

Sinha, C. (1988). *Language and Representation. A Socio-naturalistic Approach to Human Development.* New York: Harvester Press.

Sinha, C. (2004). The evolution of language: From signals to symbols to system . In D. Kimbrough, & U. Griebel, *Evolution of Communication Systems: A Comparative Approach* (pp. 217-235). Cambridge, MA: MIT Press.

Sinha, C. (2009). Objects in a storied world: Materiality, Normativity, Narrativity. *Journal of Consciousness Studies , 16* (6-8), pp. 167-190.

Sinha, C. (in press). Language as a biocultural niche and social institution. In V. Evans, & S. Pourcel, *New Directions in Cognitive Linguistics.* Amsterdam: John Benjamins.

Sinha, C., & Lopéz, J. d. (2000). Language, culture and the embodiment of spatial cognition. *Cognitive Linguistics , 11* (1-2), pp. 17–41.

Sonesson, G. (1989) *Pictorial concepts.* Lund: Lund University Press.

Sonesson, G. (2007a). From the meaning of embodiment to the embodiment of meaning: A study in phenomenological semiotics. In T. Ziemke, J. Zlatev, & R. Frank, *Body, Language and Mind,* 85-127. *Vol 1: Embodiment.* Berlin: Mouton de Gruyter.

Sonesson, G. (2007b). The extensions of man revisited. From primary to tertiary embodiment. In J. M. Krois, M. Rosengre, A. Steidele, & D. Westerkamp, *Embodiment in Cognition and Culture* (pp. 27-56). Amsterdam: Benjamins.

Sonesson, G. (2009). New Considerations on the Proper Study of Man - and, marginally, some other Animals. *Cognitive Semiotics , 4.*

Stern, D. N. (2000 [1985]). *The Interpersonal World of the Infant: A View from Psychoanalysis and Developmental Psychology.* New York: Basic Books.
Stjernfelt, F. (2007). *Diagrammatology.* Dordrecht: Springer.
Taglialatela, J.P., Russell, J.L., Schaeffer, J.A. and Hopkins, W.D. (2009). Visualizing vocal perception in the chimpanzee brain. *Cerebral Cortex* 19(5):1151-1157.
Tomasello, M. (1999). *The Cultural Origins of Human Cognition.* Cambridge, Mass.: Harvard University Press.
Tomasello, M. (2003). *Constructing a Language. A Usage-based Theory of Language of Language Acquisition.* Cambridge, MA: Harvard University Press.
Tomasello, M. (2008). *Origins of Human Communication.* Cambridge, Mass.: MIT Press.
Thompson, E. (2007). *Mind in Life: Biology, Phenomenology and the Sciences of Mind.* Cambridge, MA: Harvard University Press.
Trevarthen, C. 1979. Communication and cooperation in early infancy: A description of primary intersubjectivity. In M. Bullowa (Ed.), *Before speech,* 321-347. Cambridge: Cambridge University Press.
Trevarthen, C. and Hubley, P. (1978). Secondary intersubjectivity: Confidence, confiding and acts of meaning in the first year. In A. Lock (Ed.), *Action, Gesture and Symbol: The Emergence of Language,* 183-229. London: Academic Press.
Tylén, K. (2008). *Roses, Icebergs, Hoovers and all that Language. An investigation of the cognitive foundations of our comprehension of object mediated communication.* Ph.D Thesis: Odense: University of Southern Denmark.
Varela, F., Thompson, E., & Rosch, E. (1991). *The Embodied Mind.* Cambridge, Mass.: The MIT Press.
von Uexküll, J. (1982 [1940]). The theory of meaning. *Semiotica , 42* (1), pp. 25-82.
Vygotsky, L.S. (1962 [1934]). *Thought and Language.* Cambridge, Mass.: The MIT Press.
Wallace, B., Ross, A., Davies, J., & Anderson, T. (2007). *The Mind, the Body and the World. Psychology after Cognitivism?* Exeter, UK: Imprint Academic.
Wich, S.A., Swartz, K.B., Hardus, M.E., Lameira, A.R., Stromberg, E. and Shumaker, R.W. (2009). A case of spontaneous acquisition of a human sound by an orangutan. *Primates* 50:56–64.
Wilson, M. (2002). Six views of embodied cognition. *Psychonomic Bulletin and Review , 12* (4), pp. 625-636.
Wittgenstein, L. (1953). *Philosophical Investigations.* Oxford: Basil Blackwell.
Woll, B. and Kyle, J.(2004). Sign language. *Encyclopedia of language and linguistics.* Oxford: Elsevier.
Zahavi, D. (2003). *Husserl's Phenomenology.* Stanford: Stanford University Press.
Ziemke, T. (2003). What's that thing called embodiment? In R. Alterman, & D. Kirch, *Proceedings of the 25th Annual Meeting of the Cognitive Science Society* (pp. 1305-1310). Mahwah, NJ: Lawrence Erlbaum.
Ziemke, T, Zlatev, J. & Frank, R. (2007) (Eds.) *Body, Language and Mind. Vol 1: Embodiment.* Berlin: Mouton de Gruyter.
Zlatev, J. (1997). *Situated Embodiment: Studies in the Emergence of Spatial Meaning.* Stockholm: Gotab.
Zlatev, J. (2003). Meaning = Life + (Culture): An outline of a unified biocultural theory of meaning. *Evolution of Communication , 4* (2), pp. 253-296.
Zlatev, J. (2005). What's in a schema? Bodily mimesis and the grounding of language. In B. Hampe, *From Perception to Meaning: Image Schemas in Cognitive Linguistics* (pp. 313-343). Berlin: Mouton de Gruyter.
Zlatev, J. (2007). Embodiment, language and mimesis. In T. Ziemke, J. Zlatev, & R. Frank, *Body, Language and Mind. Vol 1: Embodiment,* 297-337. Berlin: Mouton de Gruyter.
Zlatev, J. (2008a). The co-evolution of intersubjectivity and bodily mimesis. In J. Zlatev, T. Racine, C. Sinha, & E. Itkonen, *The Shared Mind: Perspectives on Intersubjectivity,* 215-244. Amsterdam: Benjamins.
Zlatev, J. (2008b). From proto-mimesis to language: Evidence from primatology and social neuroscience. *Journal of Physiology, Paris,* 102: 137-152.
Zlatev, J. (2008c). The dependence of language on consciousness. *Journal of Consciousness Studies , 15* (6), pp. 34-62.
Zlatev, J. (2009). The Semiotic Hierarchy: Life, Consciousness, Signs, Language. *Cognitive Semiotics , 4*: 169-200
Zlatev, J. (in press). Cognitive linguistics and phenomenology and D. Schmicking and S. Gallagher (Eds.), *Handbook of Phenomenology and Cognitive Sciences.* Dordrecht and New York: Springer, 2009.
Zlatev, J., & Andrén, M. (2009). Stages and transitions in children's semiotic development, XX-XX. In J. Zlatev, M. Andrén, M. Johansson-Falck, & C. Lundmark, *Studies in Language and Cognition.* Newcastle: Cambridge Scholars Publishing.
Zlatev, J., & SEDSU-Project. (2006). Stages in the Evolution and Development of Sign Use (SEDSU). In A. Cangelosi, A. D. Smith, & K. Smith, *The Evolution of Language: Proceedings of the 6th International Conference (EVOLANG6)* (pp. 379-388). New Jersey: World Scientific.
Zlatev, J., Racine, T., Sinha, C., & Itkonen, E. (2008) (Eds.). *The Shared Mind: Perspectives on Intersubjectivity.* Amsterdam/Philadelphia: Benjamins.

Cybernetics and Human Knowing. Vol. 16, nos. 3-4, pp. 175-186

A (Cybernetic) Musing: Design and Cybernetics

Ranulph Glanville[1]

Introduction

In 1969, Gordon Pask published a paper that explicitly proposed a vital connection between cybernetics and architecture. "The Architectural Relevance of Cybernetics" (Pask, 1969) was one outcome of an extraordinary series of debates and presentations centred around the theme of limits to science presented at the Architectural Association School of Architecture (AA) in London. The proceedings were published in the journal *Architectural Design,* at the time one of the most open and experimental publications in the UK. I give these details to show how open the AA in particular, and architecture in general, were to other subjects. At the time, Pask had been on the staff of the AA as a consultant for several years. He not only argued the architectural relevance of cybernetics, he lived in an environment where this was accepted and acted upon.

Pask's central argument concerned conversation. Three years before he published officially on *conversation theory,* he explained how conversational exchange could help client and architect develop a proposal that became better than it would have been if simply briefed by initial instruction. Fourteen years later, Donald Schön (see Schön, 1983), a professor of planning at MIT with an interest in education and systems, examined the knowledge professionals develop and use in the practice of their professions. He referred to this as *reflective practice.* His insights were taken up by architects (one of the professions he examined) and other designers. He also examined the environment in which architects and designers are educated and work: the studio.[2] Schön borrowed the idea of conversation (a reflexive conversation with the situation) to explain the central act of the designer: holding a conversation with oneself through paper and pencil. This was not a new insight: many teachers of architecture, including Pask and myself, were using this metaphor—a metaphor presented to me when I was a student (What is the drawing telling you?).

So it can be seen there is reason for assuming a critical connection between cybernetics and architecture and design. I have alluded to such a relationship in these columns. I hold we can consider design as a practical expression of cybernetics, cybernetics as a theoretical study sustaining design. I have never presented the argument coherently in this column. I aim to make good this failing, here.

---

1. CybernEthics Research, Southsea, UK, email ranulph@glanville.co.uk
2. I have written about studio education in this journal: see Glanville (2003). See also Schön's study commissioned by the Royal Institute of British Architects, Schön (1985) and Broadbent et al.'s (1997) follow up.

The Shape of the Column

Writing in *Cybernetics & Human Knowing*, I assume we share an understand of what cybernetics is. If I am wrong, there are many articles readers can refer to.

Design is a different matter, not least because those who originally used the word in English use it in one way, while a number of other, later, users use it in a different, restricted manner. So I start with an essay on design as I would like it to be understood in this column, especially highlighting the importance of delight, and (what I take to be) the central act of design—which generates novelty and assimilates and accommodates complexity, and which I elaborate on below. Having rehearsed something of my understanding, I explain how design and cybernetics mirror one another.

Design

It is generally agreed that the first western text on architecture and design, dating from around the year 0, is Vitruvius's *De Architettura*.[3] Vitruvius states that architecture[4] (within which I include design) is constituted of three parts, normally arranged in a triangle:

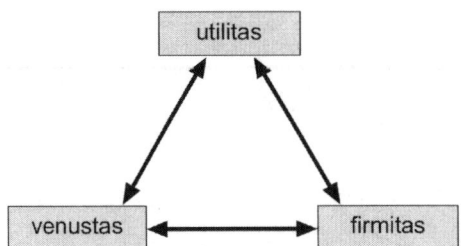

Vitruvius's Latin text was first translated into English in 1624 by Sir Henry Wotton, who used the terms *firmness*, *commodity* and *delight* ("Well building hath three conditions: firmness, commodity, and delight").[5] Contemporary usage might be well-constructed, fit-for-purpose, and delightful, sometimes stated as fabrication, function and form.

---

3. Retrieved October 15, 2009 from http://en.wikipedia.org/wiki/Vitruvius
4. Leandro Mazaro reminds me Vitruvius stated his understanding of the scope of architecture, thus: "There are three departments of architecture: the art of building, the making of time-pieces, and the construction of machinery. Building is, in its turn, divided...." I include this here to remind the reader that architecture is not uniquely identified with building.
5. Retrieved October 15, 2009 from http://en.wikipedia.org/wiki/Henry_Wotton

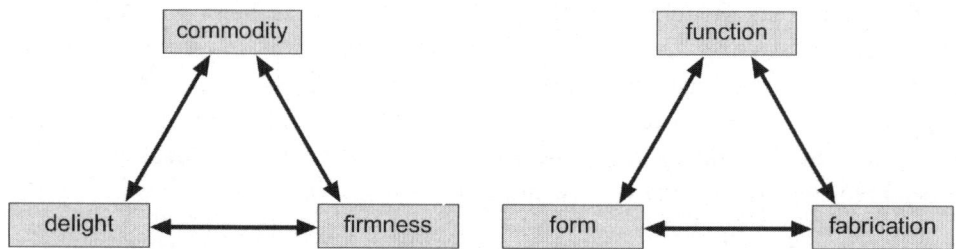

Fitness-for-purpose and function are relatively easy to specify and test for.[6] Delight, being harder, is often left out, often with the excuse that it is unscientific. And then, whose delight are we considering? In my view, delight is for all. That includes the client, users, constructors, and designers—delight both in the object or process produced, and in the designing of it. (In English, the word *design* takes the form of both noun and verb. I am primarily interested in design as verb.)

This difficulty, of accommodating delight, can be read as a commentary on the modernist slogan, first stated by American architect Louis Sullivan (inventor of the skyscraper). He insisted:

Form follows function.[7]

This slogan became a credo amongst designers of the *modern movement*, and amongst the design theorists of the immediate post-second world war period. It was taken to imply that if you dealt properly with functional requirements, the appropriate form of the designed object would arise automatically, and would bring delight. Delight could thus be ignored as a criterion: the designer need not worry about form (which, largely, gives rise to delight) because fulfilling the function generates it automatically. Other slogans reinforced Sullivan's: for instance, Le Corbusier told us the house is

A machine for living.

Early researchers into design as a field (perhaps beginning with the work in the 1950s at Hochschule für Gestaltung[8] in Ulm by Rittel[9] and his colleagues) were

---

6. Nevertheless, briefing for a new project is often difficult, specially because of our lack of practice (as Pask's 1969 paper indicates). What is needed, and how to satisfy this, is hard to capture. So are apparently minor functions (using the w.c. as a bolt hole because it is lockable, for example).

7. In architecture and design, the word form is used to refer to shape, balance, composition and symmetry, rather than to the abstract forms of (for instance) Platonic and Kantian philosophy.

8. It is interesting to consider words used for design in various European languages. The German refers to the whole (*gestaltung*, as in gestalt psychology). The Dutch use *vormgeving* (to give form to). The Italians use *disignare* (to draw, to designate or give significance to), which relates to the Greek, σχεδιάζω (in the Latin alphabet, *skediaze*, to intend or sketch).

9. Rittel later modified his position, introducing the concept of the wicked problem—Rittel and Webber (1974).

trapped within positivist science. They tended to consider design not so much as they found it, but as they thought it should be. A general view at the time was that design was defective science. Therefore, researchers concentrated on replacing what designers did by a new, scientific approach. Given delight was difficult to formalise, and the slogans (above) made delight arising from form-giving a guaranteed consequence of good functioning, it was not hard for researchers to deal only with the simply characterisable: construction and, especially, function.

The exclusion of delight is particularly apparent in Engineering Design, where it is often seen as superfluous—a trivial distraction. Engineering Design comes out of Engineering, attaching the notion of design to itself without apparent concern for how the word came into English or the (designers') original usage. I hazard a guess that the word design was added to increase status.

I would not want to suggest engineers never produce the delightful. The best engineers are wonderful designers (in the original sense of the word), producing work of imagination, quality and delight. Nevertheless, I have rarely found reference to delight, in engineering design publications. If some of the objects produced are delightful—as when delight arises out of form following function—that is an incidental by-product. Yet, more than half of the total publications published as design research come from this stream,[10] so design is often portrayed without that crucial concern (delight): those who acquired the term design have started to re-define it to suit their particular, and (in my view) constrained view.

Engineering design pursues a broadly scientific approach. The anticipated benefit is that requirements to be satisfied can be specified and quantified. The problem is reduced to atomic components, assembled in a logical manner, generating an unquestionable result. I maintain this approach is the opposite to the approach that typifies the original users (in English) of the word design, denying its central act. I will explore what this act is and how we use it to our benefit, further on. For now, it is enough to recognise that there is a difference.[11]

The significance of delight in design finds expression in another aspect. Design is about doing more than simply satisfying the necessary (being well-built and fit-for-purpose). Consider this statement attributed to the architect Sir Denys Lasdun, who held

Our job is to give the client not what he wanted, but what he never knew he wanted till he saw it.[12]

---

10. I owe this information to a conversation I had with Dr Terry Love, October 2008.
11. There is another confusion, between design as discussed in this paper and design meaning fashion/style. This "designer" notion is not our concern. However there is a profound and subtle understanding relating to design-as-style (paralleling learning styles), which is (personal) style as a way of believing complex problems are solvable. I owe this insight to Prof Stephen Gage.
12. I have had this quote in my database for several years, with the annotation that it appears in Lasdun's *Times* obituary. Unfortunately I have not been able to confirm this recently. Having known Lasdun, I am sure it is correct. It is certainly a quote that reflects accurately the view many architects hold, which can appear arrogant but is, I believe, based firmly in the notion of humble service.

This statement insists the architect/designer should strive to do more than satisfy requirements, give more than the necessary. This is an act of generosity. The concept of generosity sits well with delight: it is delightful, as giving delight is generous. Designers try to put a bit extra in, always aiming to achieve more than the strictly necessary. In the case of architects this is rightly so: In order to create what they create, they generally destroy something first (their building has to go somewhere): There is an implicit ethical requirement that they produce something better than what they have taken away.[13]

.   As a researcher into and teacher of design, my interest is in the difficult stuff: how those who do design can understand their doing in a way that empowers, coupled with an insistence on the value of delight. I hold design research that fails to consider this is inadequate: a form of research in which what designers do is seen as material for other approaches to exploit, rather than as a source of a type of research and generation of knowing that comes from and is sensitive to the subject itself.

The Act at the Centre of Design

I have presented a short account of (the use of the word) *design*, and have characterised an approach that, producing a type of scientific research, fails to recognise both the original users' use of the word, and one of the criteria for design presented in Vitruvius's original text—delight. I place so much emphasis on delight because it is the one of Vitruvius' three parts that is downplayed—and wrongly so! Now I should elaborate on what I have come to understand as the central act of design. I have argued that aspects of what the designer does are relatively straightforward: functional requirements should be satisfied, the outcome should be fit for purpose, and well-enough constructed. Often challenging and always important, these are not at the heart of my concern, which is how the designer comes up with what he/she does come up with. This is the aspect I shall deal with. If I don't mention the other aspects (Vitruvius's *firmitas* and *comoditas*) it is not because they are not important: I take them as given. But they are not what makes design special—they are not associated with a particular behaviour leading to a unique outcome.

In my account, the central act of design involves the designer holding a conversation with him/herself through the medium of paper and pencil in an act of doodling or sketching.[14] (I don't mean exclusively or literally paper and pencil, but to indicate a simple way of making marks.)

---

13. This way of describing the situation in which creation of architecture takes place, which he calls *KillSpace*, is due to the Belgian architect and media artist, Marc Godts.
14. According to the Apple Oxford Dictionary of American English, the word *sketch* comes into English in the mid 17th century from the Greek *skhedios,* meaning done extempore, probably via the form of the Dutch *schets* (see, also, footnote 7). *Doodle*, meaning to scribble absentmindedly, entered English in the early 17th century from the German *dudeltopf*, meaning simpleton. The current sense of the absent minded doodle developed in the 1930s. Some people are very proprietorial about their personal use of these words and the distinctions they make, to the point that friendships are lost! I use them more or less interchangeably, though for me doodle is a bit more mindless and a bit less purposeful (and more playful) than sketch.

To understand how this works as a mechanism, I will use another idea Gordon Pask (1975) clarified, which I suspect most readers will recognise without difficulty. In his conversation theory (developed from a formalisation of exchange mechanisms in normal conversation), Pask differentiated between psychological individuals (abbreviated p-ind's) that carry out mentative processes, and their embodiments in mechanical individuals (m-ind's) which house them. This differentiation allows for phenomena such as group intelligence, in which what may be construed as one intelligence is shared in/across many (separate) mechanical individuals (bodies). It also permits more than one p-ind within a single m-ind (body). In this latter case, Pask's formulation allows different personae to dominate at different times, or to co-exist so more that one persona is present simultaneously.

Pask does not argue humans suffer from multiple personality disorder. He points to what many of us realise: On different occasions, we behave in different ways, as if we were different people. So, for many of us, talking and listening require the assumption of different personae: we might think of the talker leading, the listener following. When I switch from talking to listening, I switch not only what I'm doing, but aspects of who I am (the role I'm taking). We recognise our ability to assume different personae in expressions such as "wearing my cybernetician's hat."

The designer, sketching or doodling, switches between the roles (personae) of marker and viewer—or, to be pedantically precise, the drawer-who-then-listens-before-drawing-again, and the listener-who-then-draws-before-listening-again (but I shall restrain myself to the simpler drawer, of I-the-drawer and listener, of I-the-listener)—a visual equivalent of talking and listening. The mark is often made without intention: it's not the shape of something, it's an exploration, a vague question. Make a mark, view it, remake (change) the mark, review it. This is a type of play, full of unspoken "What if?" questions, the form of a conversation held with oneself: statement uttered, statement heard, statement restated.

The point of a conversation is that it allows communication between personae (p-ind's) that construe the world differently. It does not presume meaning is communicated: rather, each persona constructs its understanding (hence, meaning), allowing it to behave in concert with its partners-in-conversation. Within the same body, I-the-drawer and I-the-viewer, seeing differently what is taken to be the same (the marks), offer insights to their partner participant that are different, through this mismatch, from what was previously understood. In other words, personae create novelty for/with each other: Sketching/doodling leads, inevitably, to change. The designer, sketching/doodling, starts somewhere but ends somewhere else, often unable to explain the move from the one place to the other.

A possibly mythical story offers a powerful example. The painter Wassily Kandinsky is commonly thought to have invented abstract art. One day, looking down his studio, he saw some paintings he did not recognise. He could not work out who the artist was and what the paintings represented. On closer inspection he realised the paintings were his, but placed upside down. When he painted them, he understood them one way: Returning, he saw them differently, thus construing them anew, inventing abstract art in the process.

I maintain this circular[15] act of conversing with oneself (normally through a medium such as paper and pencil), with the concomitant switch between personae (often achieved so fast that both effectively co-exist), is the central activity in designing. I have argued it is fundamental to how we behave, and may be seen as the origin of cognitive activity (Glanville, 2006b).

The reader might ask what evidence there is for my assertion. I will answer in two ways. First, recognition. When I give this account of what they do to designers, the response is recognition. They understand my point, which resonates with their reflections on their experience. This is not scientific evidence, but it is strong evidence gathered in a manner which reflects the sort of knowledge Donald Schön (1983) claimed is at the heart of professional activity and knowledge acquisition: reflective practice. It is recognised by professionals such as architects in exactly this self-referential manner: acting through reflection.

Second, a growing body of work argues both from principle and from experimental and observational work. Although designers might attribute the original conceptualisation of design as a conversation with oneself through drawing with Schön (in his work on the architectural studio, 1985), Pask introduced the conversation in his 1969 paper. I am sure we could find others who precede Pask. This aside, there is a recent burgeoning of work by a number of scholars characterising design as conversation, summarised in Lawson (2004). Henrik Gedenryd's (1988) doctorate *How Designers Work* is one of the most sensitive and revealing studies. More recently, Alice Lo (2008) has compiled and edited a book of case studies in personal design processes by staff/student pairs, which are essentially conversational. There are many other examples. Pask and I talked of the central design act as conversational already when I was his cybernetics student in the early 1970s. In other words, there is a body of scholarly work that supports my position.

Cybernetics and Design

I now argue some connections between design and cybernetics, using the notion of design conversation developed earlier. I am neither the first, nor the only person who believes there is a link between cybernetics and design: A 2007 double issue of *Kybernetes* I edited (Glanville 2007a)[16] contained 27 papers on the theme.[17]

Conversation demonstrates several key cybernetic concepts. It is circular and iterative: a feedback loop. Two (or more) participants each hear what the other has to say, and repeat back their understanding to the other, in their own formulation. X talks to Y; Y listens, constructs his understanding and talks to X. X, in turn, listens and constructs her understanding. Differences (errors) may be corrected by comparing

---

15. Some would prefer to say spiral. The progress of the conversation may be spiral, but the form within which this progress occurs is circular: Hence my choice to describe it that way.
16. With assistance from Ben Sweeting.

understandings before and after this exchange. Organisationally, a conversation is essentially the same as a thermostat: The difference is in how we understand the enhanced mentative abilities of the elements X and Y.

A diagrammatic presentation of the stages in a normal Paskian verbal conversation may help:

Figure 1

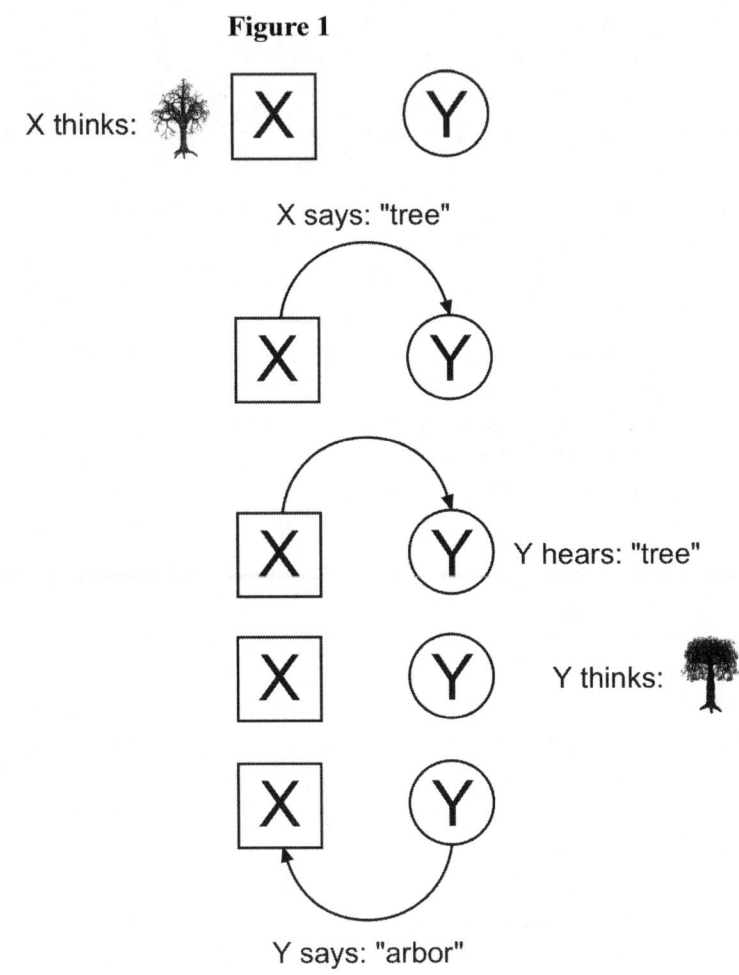

Figure continued next page

Figure continued next page

---

17. One author was Klaus Krippendorff. Krippendorff was educated as a designer in Ulm (1956–61), and studied cybernetics under Ashby. He and I thus share a background, although I think we have radically different interests and hold radically different positions about the relationship between the individual and society (Krippendorff sees the individual residing in society, I see society growing out of aggregations of individuals). I wrote a review article of his book *The Semantic Turn* (Krippendorff, 2006) in this journal (Glanville, 2007b), concentrating on his development of a way of handling *the user* rather than his *science for design* because of this difference, and particularly its consequences when we consider the source of meaning and the significance of language, which Krippendorff holds to be primary, but which I do not. Here, I recognise aspects of a common background, but note Krippendorff's interest in design, and mine, are very different: and that my interest lies in what I have referred to as the act at the centre of design. My wish is to explain my position, not to argue against his: but, because of the similarity of background, I felt I had to mention his work.

Figure 1, cont'd

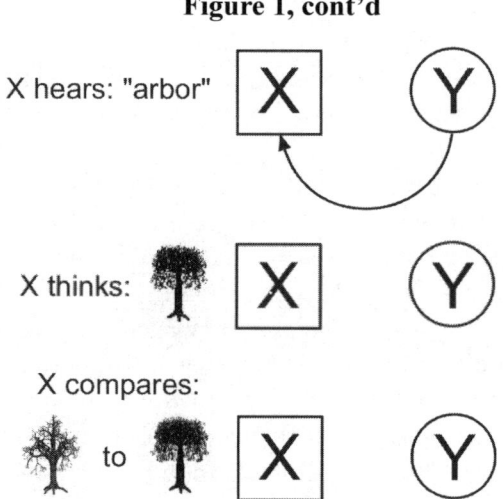

It should be noted that, in a normal conversation, the comparison at the final step may be used to minimise difference, whereas in design it is often used to enhance, or at least accept the difference.

The design conversation (held with the self in a different role) is a modification of the basic conversational form, in that X and Y are often within one body, understood as different personae, rather than different people; and the utterances are mainly drawn and viewed rather than said and heard. However, whereas in most models of communication the concern is to reduce error, in design the so-called "error" may be a source of novelty. What is often thought of as error is welcomed as a means of enhancing creativity.

This novelty comes from everything in the system working together. Ross Ashby (1991) explained that the description of even such an apparently mechanistic device as the Black Box arises from interaction between Black Box and its investigator (Glanville, 2009). Gordon Pask understood this from his earliest work. The outcome of this behavioural interaction is unpredictable and beyond what can be achieved by one participant alone. Unfortunately, interaction has been reduced to responsiveness by the computer industry, to the impoverishment of our conceptual world. Interaction is behind the novelty generated. Conversation epitomises interaction, perhaps its simplest, clearest and best formalised example.

Other important behaviours of this essentially cybernetic act are as a means to encourage and manage accommodation (the process of adapting and adjusting to someone or something else), affordance (in the opportunistic sense of J. J. Gibson, 1979: finding opportunity in objects and processes that were not part of an original intention) and assimilation (to absorb, integrate and fully understand ideas). In a conversation, that which was not expected may arise (named as novelty and creativity), and be taken on board. But we can also bring in, rather than generate, that which was not previously under consideration. Thus, we accommodate needs not previously considered; see what we have done in a new light affording unanticipated possibilities; and assimilate separated concepts which become integral. These three moves (essential to the development of our cognitive abilities) are crucial to the development of a design scheme. They are not (in my view) central design acts, but ways we can use that central activity to help us deal with design tasks and enhance our design ability.

These connections are not the only possibilities: they reflect my interests, establishing a metaphor with key understandings in second order cybernetics.

Many people demonstrate other types of connection in architecture: the cybernetic control of systems within buildings (lifts and heating systems); systems that change the building (responsive louvres); cybernetic management principles (ordering the construction of the building); the cybernetics of communication (between those involved); buildings that result from cybernetic actions (in some form of automatic generation or as a consequence of a cybernetic act, for instance space stations); even the image of cybernetics (eg., the work of Archigram).

The list goes on. I make no attempt to complete it.

Design and Cybernetics

Hidden in the above, is an unexpected (and novel) extension of cybernetics.

Describing the central act of design through the metaphor of conversing with oneself connects to the idea different personae don't see the world in the same way, form the same understandings, or know the same things. We can reconstrue this: The variety of any one persona cannot equal the variety of all (other) personae. Readers may remember when I wrote of unmanageability as a way to enhance creativity (Glanville, 1998). Normally, cybernetics is interested in systems which conform to its one universally accepted law, Ashby's law of requisite variety, thus being manageable. In contrast, I propose we should develop an interest in the unmanageable: a form of anti-cybernetics.

So what about doodling? I described it as purposeless. If doodling has a purpose, it may be to find (rather than assume) purpose. One of the basic assumptions of cybernetics is that we deal with purposive systems: Wiener's first cybernetic paper (Rosenblueth, Wiener and Bigelow, 1943) is notable for bringing purpose (intention/goal) into scientific discussion.[18] It may, now, be important to consider a cybernetics of the purposeless.

Thus, I position myself as much as an anti-cybernetician as a cybernetician, much in the manner I value ignorance as much as I value knowing (for ignorance is a source of knowing). Is not the unmanageable and the purposeless equally the source of variety and purpose, of cybernetics? If so, telling the story with design as the conceptual source, expands and enhances cybernetics, as our understanding of cybernetics helped us account for design.

---

18. Of course, we have much older examples: What would Darwin's theory of evolution be without purpose?

Conclusion

I have spent a major part of this column discussing how I understand design, why some uses of the word might be considered inappropriate, the particular criterion of delight and the importance of form, for I believe readers of this journal are not particularly familiar with design, at least in the understanding used here. I characterised an activity I hold is at the heart of design, and how this activity—the conversation with oneself via paper and pencil—is so very cybernetic: and I extended it to include the mechanisms of accommodation, affordance and assimilation.

This is why I claim cybernetics may be thought of as the theoretical arm of design while design may be thought of as the practical arm of cybernetics.

I can summarise the position argued, quoting from a talk I gave in 2006 (Glanville, 2006a):

- Design, according to Vitruvius, deals with three qualities: firmitas, utulitas and venustas.
- Conversation is essentially constructivist: each participant constructs his/her own meaning and value (therefore, each is responsible for this).
- Design is a conversation held primarily with the self (but also others): self-conversation emphasises the significance of listening/being receptive.
- Designers develop and amplify ideas, make the new from differences in meanings—when difference in expression is welcomed, not hidden.
- The process of design is circular, iterative, unknowing (including rejecting and restarting), constructive: explanations are post-rationalised.
- The new is beyond prediction.

Implicit in conversation (and thus design) are many ethical qualities we think of as deeply human and desirable.

References

Ashby, W. R. (1991). General systems theory as a new discipline. In G. J. Klir, *Facets of systems science* (pp. 249–257). New York: Plenum Press. (Originally published in 1958)

Broadbent, G., Martinez, A., Cardaci, E., & Zoilo, A. (1997). The design studio revisited. *Environments by Design, 2* (1).

Gedenryd, H. (1998). How designers work: Making sense of authentic cognitive activity. *Lund University Cognitive Studies* [No.] 75. Lund, Sweden: Lund University. Retrieved October 11, 2009 from http://web.archive.org/web/20021029071724/http://lucs.fil.lu.se/People/Henrik.Gedenryd/HowDesignersWork/index.html

Gibson, J. J. (1979). *The ecological approach to visual perception*. Boston: Houghton Mifflin.

Glanville, R. (1998). A (Cybernetic) Musing: Variety and creativity. Less than the requisite variety—or the wonder of the unmanageable in art, teaching and the world wide web. Cybernetics & Human Knowing, 5 (3).

Glanville, R. (2003). A (cybernetic) musing: Some examples of cybernetically informed educational practice. *Cybernetics & Human Knowing, 9* (3).

Glanville, R. (2006a). *Design and construction*. Keynote address presented at the Heinz von Foerster Conference, Vienna, November 14, 2006.

Glanville, R. (2006b). Design and mentation: Piaget's constant objects. *The (Radical) Designist*. Retrieved October 11, 2009 from www.iade.pt/designist)

Glanville, R. (Ed.) (2007a). Cybernetics and design (special issue). *Kybernetes, 36* (9-10).

Glanville, R. (2007b). Review: Design, the user, and Klaus Krippendorff's "The semantic turn." *Cybernetics & Human Knowing, 14* (4).

Glanville, R. (2009). Black boxes. *Cybernetics & Human Knowing, 16* (1–2).

Krippendorff, K. (2006). The semantic turn: A new foundation for design, Boca Raton, FL: CRC Press.

Lawson, B. (2004). *What designers know.* Oxford: Architectural Press.

Lo, A. (Ed.) (2009). *We made it!* Hong Kong: Hong Kong Polytechnic University School of Design.

Pask, G. (1969). The architectural relevance of cybernetics. *Architectural Design*, (9).

Pask, G. (1975). *Conversation theory.* London: Hutchinson.

Rittel, H., & Webber, M. (1974).Dilemmas in a general theory of planning. *Policy Studies, 4*, 155–168.

Rosenblueth, A., Wiener, N., & Bigelow, J. (1943). Behavior, purpose and teleology. *Philosophy of Science, 10* (1), 18-24.

Schön, D. (1983). The reflective practitioner: How professions think in action. London: Basic Books.

Schön, D. (1985). The design studio: An exploration of its traditions and potentials. London: RIBA Publications for RIBA Building Industry Trust.

Craft, D. (2009). *Bardo*. Microphoto: Carbolic Acid Melt Crystals; 34 x 34 cm.

Cybernetics and Human Knowing. Vol. 16, nos. 3-4, pp. 187-194

ASC
American Society for Cybernetics
a society for the art and
science of human understanding

Bio-cost
An Economics of Human Behavior

Hugh Dubberly,[1] CJ Maupin[2] and Paul Pangaro[3]

Much of human behavior is directed toward goals: finding food, selling services, curing cancer, making meaning.

Achieving goals requires action. Action requires effort. Effort requires energy and attention applied over time. Effort overcomes obstacles. Obstacles tax our patience, sap our resolve, and cause us stress.

English (as well as many other languages) includes many metaphors that frame effort as a cost:

- I enjoy **spending** time with you.
- You're **wasting** your energy.
- You're not **paying** attention.
- This job is not **worth** the stress.
- It all takes its **toll**.

These metaphors suggest an economics of human behavior—a framework for understanding the human cost of living and the trade-offs we make moment-by-moment as we choose one course of action over another. This paper begins the development of such a framework for everyday living and suggests how it might be applied to business and design. The authors hope to provide a means for us all to learn to act in better accord with our interests and thereby improve productivity and satisfaction, both individually and in concert with others.

Bio-cost measures human effort

Bio-cost is the energy, attention, and stress that people expend over time to achieve their goals—to *get what they want* in Ashby's sense (Ashby, 1956).

1. Dubberly Design Office, 2501 Harrison Street, No. 7, San Francisco, CA 94110 Email: hugh@dubberly.com
2. Cybernetic Lifestyles, 200 W 58th St, No. 5B, New York, NY 10019 Email: cj@cyberneticlifestyles.com
3. Cybernetic Lifestyles, 200 W 58th St, No. 5B, New York, NY 10019 Email: pan@pangaro.com

All of life's activities carry some bio-cost. Most often, we feel bio-cost when we meet resistance—when we can't enter a flow and act simply to get what we want. We experience the drain of bio-cost every day—when we find a stone in our shoe; when traffic slows us; when we struggle to change a channel with a remote control; when the bureaucracy requires we submit another form; when the boss makes contradictory requests; when the stock market sends mixed signals.

Bio-cost limits what we can achieve because we may not have the resources to get what we want, or we might spend too much for what we get in return. This is true for individuals, groups, organizations, and species. While we may not be able to quantify bio-cost with precise measures—whether in anticipation of expending it or after the fact—the authors have found considerable utility in construing bio-cost as comprising distinct quantitative components.

Bio-cost Is a Function of Time

All tasks take time to accomplish. The effort required to complete a task can be mapped against time (in basic cases, at least). Graphing against time we see an ebb and flow of effort—for example, walking to a destination requires relatively constant bio-cost expenditure over time, while flagging down a cab and getting in requires an initial burst of effort followed by a period of relative rest during the ride, as in figure 1.

**Figure 1: Bio-cost of physical effort to travel by taxi (dotted)
versus walking (solid).**

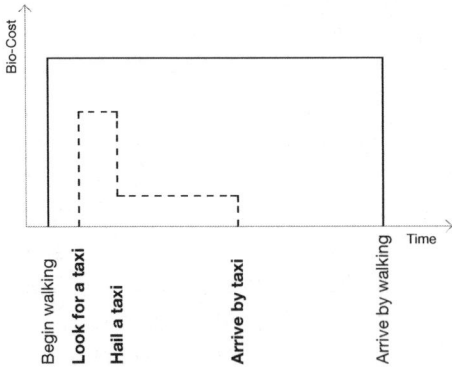

Bio-cost Has Physical, Mental and Emotional Components

In the case above, the physical effort can be measured as calories – the greater the effort, the more calories required. There are limits to our physical efforts; when taken to an extreme, we can experience muscle fatigue or exhaustion.

Bio-cost also has a mental component. Mental effort means attention paid to perform a task or even to think about how to perform it. As with physical effort, this use of our brains and all the components of our nervous system that coordinate our thinking and acting also requires effort and also has limits. Some tasks require more concentration than others, so the attention we pay will vary.

Similarly, we reach emotional limits as palpable as physical and mental ones when we get "stressed out" due to factors such as uncertainty and fear.

Bio-cost Reveals Trade-offs

Because the chemical and hormonal pathways overlay the nervous system, feeling has impact on thinking and vice versa (Von Foerster, 1974/2003). A second-order awareness of the toll that a task is taking—whether in physical, mental, or emotional terms—may add further stress or alleviate it. This becomes part of a feedback loop that helps us to estimate the bio-cost expenditure required to be successful. When the task is to *save our life*—for example, to undergo invasive surgery to remove a tumor—our stress is increased because the stakes and uncertainties are high. When there are negative consequences for not completing a task by some deadline, such as getting to the airport in time to board a flight, perceived limitations of time can contribute to stress. Even non-time threatened situations raise our stress levels: Will I get fired for that mistake? Will I pass the test? Will she like me?

By reflecting on the bio-cost of specific activities in our daily lives, we can usually make trade-offs among the components—time, energy, attention, and stress as shown in table 1—to minimize the overall cost of getting what we want or need. At any point we may also decide to spend money to lower one or more dimensions of bio-cost. (Here we note without further exploration that this has the side benefit of allowing us to calculate a monetary equivalence for bio-cost, at least in a specific context. For example, avoiding the additional time and physical effort of walking is often worth the $10 monetary cost of a taxi—plus the stress of not knowing whether we can find one in time and whether traffic will cooperate.)

Table 1: Bio-cost Components

What is the bio-cost of achieving a goal?			
Component	Expenditure	How Experienced	How Compared
time	opportunity	duration	more or fewer hours: minutes: seconds
energy	physical	work	more or fewer calories
attention	mental	focus / concentration / degree of familiarity	how many / what types of multitasking possible
stress	emotional	fear / worry / anxiety / uncertainty	higher or lower risk, higher or lower enthusiasm / motivation

Can We Replenish Our Reserves?

Clearly, we cannot recover time once spent, but given more time, we may be able to replenish our energy, our ability to concentrate, and our capacity to absorb stress.

After periods of intense activity, we often seek a better *life balance*, that is, we seek to counter-act activities that carry significant bio-cost with those that allow us to restore our physical, mental, and emotional systems. For example, we often say that we "make time" for family and friends, so that we can "recharge our batteries."

Sleep appears to restore our energy, refresh our brains, and reduce our stress such that we can use our time more efficiently and make better choices. Many other activities also fit this category, such as meditation, the pursuit of sports, crafts, and the arts, or even mastery of a skill.

How Do We Assess Bio-cost Trade-offs?

In monetary transactions we commonly consider cost versus gain. This paper argues that the same is true for actions that involve the expenditure of physical, mental, and emotional effort, and that explicit awareness of this affords us the opportunity to reflect on trade-offs and improve the choices we make.

It is important to keep in mind, however, that we can't always easily calculate the value of reducing bio-cost in monetary terms nor can we translate or commute a given valuation to other circumstances or individuals. Still, we maintain a belief in the gain and a sense of the cost, and we remain capable of generating an opinion as to what we will base our actions on right now. Put another way, we think the view is worth the climb.

In order to characterize a progression of variations of goal setting, taking action, and reaping rewards, the next set of figures start from a single participant and proceed to cover cases of cooperation and collaboration with others.

Bio-cost for Single Participant

Figure 2 draws from Pask's model of goal/action systems (Pask, 1975; Pangaro, 2003), reinterpreted such that the *goal* level (L1 in Pask's original) becomes the gain, while the *means* level (L0 in the original) becomes the cost. Per Pask, the goal-level controls the execution of procedures at the means-level, as indicated by the vertical arrow on right side. Results from execution are returned and compared to the original goal, as indicated by the line on the left with comparator sign.

Figure 2: First canonical form shows goals are achieved via separate means, where the means has a cost and achieving the goal creates a gain.

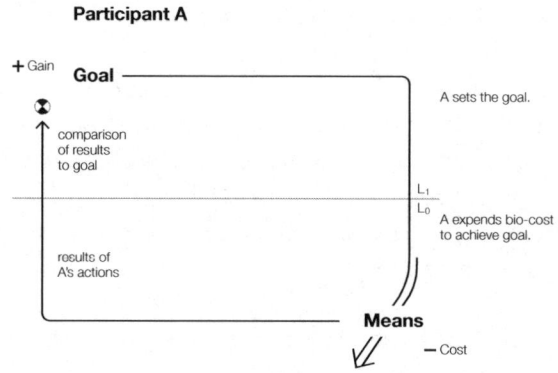

Bio-cost for Cooperative Participants

The next case involves a distinction between Participant A, who sets the goal, and Participant B, who agrees to perform the actions required to achieve that goal. The components are the same as in Figure 2. However, in Figure 3, there is a clear division (the vertical line) as goal-setting and action-taking are executed by different participants.

Participant B expends the bio-cost to achieve the goal on behalf of Participant A, who compares the result of B's actions with the goal. We call the interaction *cooperation* because there are clear roles and actions for A and for B—they co-operate, that is, they operate together but within agreed boundaries.

Figure 3: Second canonical form shows the allocation of goal-setting to Participant A, and action-taking to Participant B.

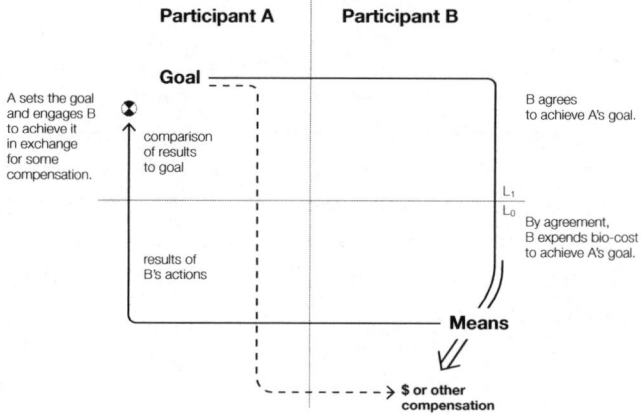

Bio-cost for Collaborative Participants

The third case also involves two participants but is more open-ended in that the distribution of roles and actions between participants is not predetermined. Rather, participants A and B collaborate – they "co-labor" or work together – to create and agree on the goals themselves, as well as to agree on who does what to achieve them.

In figure 4, participants A and B converse at two levels: about goals (upper horizontal loop) and about the means to achieve them (lower horizontal loop). They likely also cooperate about means, and use feedback to check whether goals have been achieved (loops that cross from upper to lower level). In an ongoing collaboration, participants may maintain some sense of the trade-offs across time and situations, and they may seek a balance over time.

Figure 4: Third canonical form shows that A and B "co-labor" to create goals and share bio-cost to achieve them.

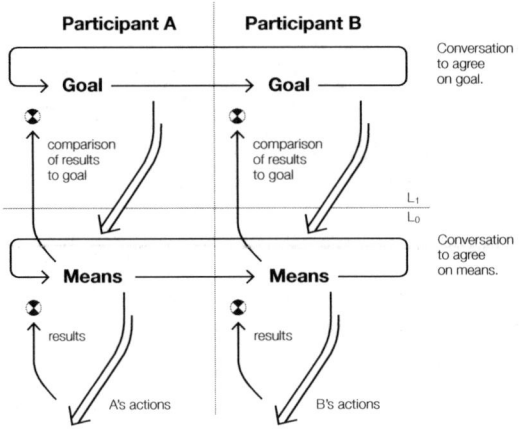

Bio-cost in Business and Design

Society has benefitted greatly from—one could say society can arise because of—the sharing of bio-cost. As early as the Stone Age, social groups learned how coordinated action could achieve goals that would otherwise have been impossible. A group could successfully hunt a swift and powerful animal for food, whereas a single hunter might have only a slim chance of success and a high risk of injury or death. By sharing such responsibilities, groups could achieve net bio-cost reduction thereby freeing up resources to explore new lands, create new arts and cultures, and develop new means of associating and collaborating.

Since the Renaissance, the corporation has provided one such structure for collaboration. The success of modern corporations is a measure of the huge scale on which they reduce collective bio-cost expenditures. Yet, modern corporations also exact a huge toll in frustration and stress from their employees. In other words, working in a corporation often comes with a high bio-cost. For example, on a mundane level the noise and interruptions of "cubicle life" can make focused attention difficult. On a more critical level, uncertainty about goals and criteria can lead to rework; uncertainty about roles and responsibilities can lead to unproductive conflict; and uncertainty about continued employment can lead to fear. Such bio-costs are an extraordinary and persistent waste of human resources.

Transforming a corporation from a current state of high bio-cost to a more efficient state requires a complex system that learns as it goes—and the bio-cost of learning, even for those who thrive on it, is very high (Geoghegan & Pangaro, 2009). This appears to be one reason why corporations often fail to find new paths to success when markets change (Dubberly, Esmonde, Geoghegan & Pangaro, 2002).

On the other hand, strong teamwork means that there is mutual trust (itself a huge bio-cost reducer) as well as clarity of direction, role and proper action (all proxies for low uncertainty and hence low bio-cost situations). At best, the beliefs and goals of the individuals in a corporation are highly aligned.

In addition to applying the framework of bio-cost to organizational design, we can also apply it to product and service design. Minimizing or at least reducing a user's bio-cost can be an important design goal. Even though the precision of bio-cost measures is limited, a focus on bio-cost permits a deep conversation during the design process. Instead of seeking to make products "simple" or "intuitive"—laudable goals but not very specific—designers can use the dimensions of bio-cost to participate in a more directed design process where trade-offs are made explicit and clear.

Why Bio-cost Is Important

We see an opportunity for organizations to create value by focusing on bio-cost. First, bio-cost provides a framework for improving productivity; by getting better at understanding bio-cost, we can get better at reducing it. In addition, bio-cost provides a framework for innovation; identifying bio-cost is identifying inefficiency, identifying an unmet user need, identifying an opportunity for new products and services.

In summary, it is our conviction that reducing bio-cost leads to:
- greater efficiency in achieving goals, which leads to
- greater capacity or resources in the system, which allows the cultivation of
- greater variety, which means
- greater ability to generate higher-level plans for reducing bio-cost even further, resulting in
- even lower bio-cost—a positive feedback loop and a virtuous cycle.

Reducing bio-cost creates value. It expands the space in which additional choices may be generated and evaluated. It can be an ethical motivation in the design process and lead to a more humane world. We believe that a bio-cost economy underlies all exchanges of value, and it always will, because it involves the management of the least fungible and most valuable aspect of life: how we spend our time.

References

Ashby, R. (1956). *An introduction to cybernetics*. London: Chapman & Hall.

Dubberly, H., Esmonde, P., Geoghegan, M., & Pangaro, P. (2002). *Notes on the role of leadership and language in regenerating organizations*. Santa Clara, CA: Sun Microsystems.

Geoghegan, M., & Pangaro, P. (2009). Design of a self-generating organization. In P. Asaro, & G. Klir, (Eds.), The intellectual legacy of W. Ross Ashby [Special Issue]. *The International Journal of General Systems, 38* (2), 155-173. (Paper originally presented at the Ashby Centenary Conference, Urbana, Illinois, March 5, 2004).

Pangaro, P (2003). *The architecture of conversations*. Monograph retrieved November 10, 2009 from: http://pangaro.com/L1L0/

Pask, G. (1975). Introduction to chapter on Machine Intelligence. In N. Negroponte (Ed.), *Soft architecture machines* (pp. 6-31). Boston: The MIT Press.

Von Foerster, H. (2003) . On Constructing A Reality. In *Understanding understanding: Essays on cybernetics and cognition* (pp. 211-227). New York: Springer. (Originally published in F. E. Preiser (Ed.), *Environmental design research* [Vol. 2, pp. 35-46] by Dowden, Hutchinons & Ross of Stroudberg, 1974)

Varela, F., Thompson, E., & Rosch, E. (1991). *The embodied mind: Cognitive science and human experience*. Cambridge, MA: The MIT Press.

Craft, D. (2009). *Terrain Following*. Microphoto: Potassium Acid Phthalate Precipitation Crystals; 34 x 21 cm.

Cybernetics and Human Knowing. Vol. 16, nos. 3-4, pp. 195-200

Book Review
Philosophy of Computing and Information – 5 Questions

Gordana Dodig-Crnkovic[1]

Philosophy of Computing and Information – 5 Questions. Edited by Luciano Floridi. Automatic Press / VIP, 2008, 204 pp. ISBN-10: 8792130097; ISBN-13: 978-8792130099. Contributors: Margaret A. Boden, Valentino Braitenberg, Brian Cantwell-Smith, Gregory Chaitin, Daniel C. Dennett, Keith Devlin, Fred Dretske, Hubert L. Dreyfus, Luciano Floridi, Tony Hoare, John McCarthy, John R. Searle, Aaron Sloman, Patrick Suppes, Johan van Benthem,Terry Winograd, Stephen Wolfram

> Computing and information, and their philosophy in the broad sense, play a most important scientific, technological and conceptual role in our world. ... This book collects together, for the first time, the views and experiences of some of the visionary pioneers and most influential thinkers in such a fundamental area of our intellectual development. – Luciano Floridi

This book is one among pearls in the *5 Questions Series* by Automatic Press (VIP) which presents answers on five challenging questions by leading modern thinkers, in this case within philosophy of computing and information.

The questions the editor, Luciano Floridi asked are the following:

1. Why were you initially drawn to computational and/or informational issues?
2. What example(s) from your work (or the work of others) best illustrate(s) the fruitful use of a computational and/or informational approach for foundational researches and/or applications?
3. What is the proper role of computer science and/or information science in relation to other disciplines?
4. What do you consider the most neglected topics and/or contributions in late 20th century studies of computation and/or information?
5. What are the most important open problems concerning computation and/or information and what are the prospects for progress?

Given the public interest in the eminent contributors, the answers to the question about how they got interested in the field of computing and information is both instructive and historically significant. They are highly personal and vivid reminiscences of the pioneering era and therefore hard to recapture in this review—they just have to be read the way they are told.

For the rest of the answers, I will give a short account for each of the contributors, often using their own words as illustration.

---

1. School of Innovation, Design and Engineering, Mälardalen University. Sweden.
    Email: gordana.dodig-crnkovic@mdh.se Website: http://www.idt.mdh.se/personal/gdc

As the editor points out in the introduction, the contributors had the freedom to interpret the questions and answer in the format they find suitable, which resulted in very different individual styles of responses, which also adds to the charm of the book.

Margaret Boden gives us a detailed account of computational ideas used to clarify fundamental questions about the nature of mind, with a number of valuable pointers and references. Boden declares: "My own view is that a naturalistic view must be possible, and that it is likely to be grounded in evolution" (p. 8). Boden is rightly warning against "regrettable hostility" (p. 5) between different approaches in cognitive science (symbolic, connectionist, situated, dynamical, and homeostatic)— "because all of them (and probably more) will be needed to emulate the rich space of possible minds" (p. 5).

Valentino Braitenberg emphasizes the importance of complexity, "not only in the brain but generally in living matter everywhere" (p. 16) and the ability of information which "properly understood, is fully sufficient to do away with popular dualistic schemes invoking spiritual substances distinct from anything in physics" (p. 16).

Brian Cantwell-Smith illustrates his own long journey in study of construals of computing to conclude that "in one way or other, computation involves an interaction or interplay of *meaning* and *mechanism*" (p. 31). When it comes to meaning (semantics) Smith is not so much interested in the relation between a program and a process which results from running it, but rather in the connection between that process and the task domain that the process is about, in other words, he is interested in the semantics of program semantics. Finally, Smith claims that ontology and epistemology must be reconstructed together, as a new metaphysics: "as we can now see a comprehensive theory of meaning/mechanism dialectic – involves nothing less than a full-fledged assault on constructing an appropriate metaphysics" (p. 44). For Smith, computers can help by serving as "laboratories of middling complexity" (p. 46) "in terms of which to explore issues of intentionality, embodiment, and semantics" (p. 47).

Gregory Chaitin relates information with his algorithmic complexity, Leibniz's argument about the necessity for natural laws to be simple, and knowledge as information compression. An important contribution to the field is his epistemology as information theory. "A scientific theory is only of value to the extent that it's a compression. The number of bits in the theory (considered as software) must be substantially smaller than the number of bits of empirical data that we are trying to understand/explain. Understanding is compression of information, a good explanation is a good compression" (p. 54). Chaitin is known for his pancomputationalism, the view that the universe is an output of a computational mechanism based on physical laws. Recently he has been applying his ideas about complexity to biology.

Daniel C. Dennett gives examples from his work, especially his essay "Artificial Intelligence as Philosophy and Psychology"(1978) as illustration of a possibility of demonstrating simplified working models of cognitive process. Dennett points out the view that computer science keeps cognitive science honest, as "if it weren't for the practical possibility of constructing and demonstrating simplified working models of cognitive processes, we'd still be at the hand-waving stage" (p. 58). As the unsolved problem Dennett selects the lack of solid theory of semantic information.

Keith Devlin makes distinction between information as a semantic concept and its syntactic representation. His approach is based on Barwise and Perry's situation theory. This of course has a very relevant social domain in which "the goal is not 'perfect understanding' but better (i.e. deeper, more precise, more illuminating, more useful) understanding" (p. 68). Devlin continues by concluding that "we learned more about language by seeing the extent to which real language both conforms and differs from Chomsky's mathematical descriptions" (p. 69). He expects, paraphrasing Max Planck, the new paradigm to win not by convincing the opponents but establishing itself among new generations of researchers.

Fred Dretske starts his research with epistemology, and information is of interest as a building block of knowledge. As knowledge by definition always is true, so must its constituent parts in this view also be true. Dretske adds however: "If, as I (once again) suspect, contributors to this volume mean something else by term 'information' then our answers to the questions posed will not only be different, they will be different *because* – and, perhaps, *only* because – they are understood to be answers to quite different questions." "The disagreements – and there are sure to be many – might not run very deep once the merely verbal differences are sorted out" (p. 76). Yes indeed, and this volume among others helps to make explicit the plurality of frameworks for the family of concepts that constitute our understanding of information and computing.

Hubert L. Dreyfus was in 1963 invited to evaluate Alan Newell and Herbert Simon's work on cognitive simulation. As a philosopher, he readily recognized that AI scientists were in practice turning rationalist philosophy (Hobbes, Descartes, Leibniz, Kant, Russell) into a research program of GOFAI. An intelligent (expert) system with a set of true statements and logics was used to assess the facts of the real world. Dreyfus' conclusion was that "the deep problem wasn't storing millions of facts; it was *knowing which facts were relevant* in any given situation" (p. 80). The alternative, Heideggerian/Merleau-Pontian program developed by Freeman, which was ontologically sound and unlike GOFAI capable of solving the problem of relevance had a different sort of problem: "a neurodynamic computer model would have to be given a detailed description of a body and motivations as ours if things were to count as significant for it so that it could learn to act intelligently in *our* world" (p. 87).

Luciano Floridi describes his search for "epistemology without knowing object" (p. 89) and methodological minimalism obtained in epistemology by step of adopting a more fundamental level of abstraction – information instead of traditional knowledge. Floridi characterizes computer and information sciences as "epistemic enablers" (p.93). He claims that philosophy of information is becoming our *philosophia prima*. According to Floridi, "one of the most neglected topics in late twentieth century studies of computation and information is a *philosophy of nature in the widest sense of the word* (that is in the German sense of Naturphilosophie as this was used by Schelling and Hegel)"(p. 94)—I can only agree.

Tony Hoare directs his account on the history and the future of the effort of making software error-free. He ends with the following optimist vision: "let me look forward to the day when programming error is a problem from the past; when computer programmers make fewer mistakes than engineers in any other profession" (p. 107)—we all look forward to that day. In this context it would nevertheless be interesting and highly relevant to learn more about Hoares study of process algebra—a mathematical formalism developed to describe systems in continuous interaction with the environment—increasingly important in new paradigms of computing.

John McCarthy started by comparing of the brain and computers, concluding that Newell and Simon's information processing approach revolutionized psychology, while destroying behaviorism. He advises that "philosophers need to adopt some of the practices of AI and first study simple variants of phenomena like action, knowledge, belief and context rather than only looking for the most general definitions" (p. 109), which is a view he shares with many among other book contributors.

John R. Searle is known as a critic of what he takes to be mistaken computational views of cognition such as strong AI claim that the mind is the software of the brain (which is hardware). Searle holds that "the most neglected topics in studies of computation and information that are psychologically real, that is, that are relevant to Cognitive Science, have to do with the question of how the brain actually works as a physical biological system" (p. 115). Searle insightfully welcomes the move from computational cognitive science to cognitive neuroscience.

Aaron Sloman represents the *design stance* (that constructs explanations of working systems by modeling and testing implementations) which in case of mind amounts to constructing machines with cognitive capacities. His conclusion is that informational architecture is central for understanding of cognition: "I begun [sic] to think about integrated information-processing architectures combining many different sorts of components, and that eventually led me to the design-based analysis of many other aspects of human minds and animal minds, constantly driven by the question: what sort of machine could do *that*?" (p. 118). When it comes to the future of the field, Sloman advocates his own virtual-machine approach to the mind. In his words: "Understanding the variety of types of virtual machines and the variety of ways in which virtual machines can be implemented or realized in physical machines or other virtual machines, will, I suspect, provide much matter for philosophical analysis in future years" (p. 135).

Patrick Suppes points out that brain computations on a system level are electromagnetic while on the cell level they are chemical. They are probabilistic and deeply parallel in structure. When it comes to the question of continuum versus discrete character of computational mechanism, Suppes interestingly refers to Kant's second antinomy—the theory that the whole consists of indivisible atoms whereas, in fact, none such exist—while in the question of the relationship between free will and causally bound mechanism, he refers to the Kant's third antinomy, which addresses the problem of freedom in relation to universal causality. As a most important open problem Suppes chooses the fundamental nature of space and time with an interesting remark: "But there is still a reluctance to develop what seems to be a natural isomorphism between discrete space-time and continuous space-time" (p. 154). One of the open questions is for Systems Neuroscience: how large collections of synchronized neurons are computing, with all relevant physics and chemistry.

Johan van Benthem describes John Barwise and John Perry's *situation semantics* as "a radical alternative to the ancient regime in philosophical and mathematical logic. On their view ... logic should study the information available in rich distributed environments (with both physical and human components), and the resulting information flow" (p. 160). In his view: "Statics and dynamics come together in modern logics of what may be called *intelligent interaction* – and this is no coincidence. Logic and information should take the systematic Tandem View that information can not be understood in isolation from the process which conveys and transforms it. *No information without transformation!*" (p. 161).[2] Van Benthem rightly notices cohesive force that concept of information presents: "interest in information and computation as themes cutting through old boundaries between the humanities, social, and natural sciences" (p. 162). He strongly emphasizes the interplay between statics and dynamics, information and process.[3]

Computer Science or rather Informatics in this context provides tools for representation of data together with methods of computation over them. Unlike Turing machines which are sequential computational models, more general formulations provide explicit representations of concurrency and communication such as the process algebra among others. Van Benthem concludes: "Taking biological and psychological facts seriously is not uncontroversial in logical circles, but 'Information, Computation and Cognition' may be the way to go" (p. 171).

Terry Winograd describes his own fundamental work as a "critical re-examination of the relationship between symbolic processing and the communicative workings of ordinary human language" (p. 163). Winograd, in agreement with Cantwell-Smith characterizes computer science as a mix of disciplines unlike classical sciences but with new and promising possibilities, both as models and tools opening a new world of examples and thought experiments. "In a sense, we can see 21st century ascendance of the biological sciences as a product of being able to deal with extreme complexity in a rigorous computational way" (p. 174). As the largest open problem Winograd mentions the "decoding of thought" as a "far-off but intriguing goal" (p. 174).

---

2. Compare to "no information without computation" as found in the introduction to: Dodig-Crnkovic G. and Stuart S., Eds. *Computation, Information, Cognition – The Nexus and The Liminal*, Cambridge Scholars Publishing, Cambridge, 2007

3. In my research in the field I came to the same conclusions. See Gordana Dodig-Crnkovic: *Investigations into Information Semantics and Ethics of Computing* http://www.diva-portal.org/mdh/theses/abstract.xsql?dbid=153, Mälardalen University Press, September 2006.

Stephen Wolfram spent some twenty-five years applying computational ideas to fundamental science. Together with Chaitin he is one of the most prominent representatives of pancomputationalism (natural computationalism). He explains the importance of his research: "Learning about computational universe also informs many old foundational questions in science and elsewhere. It shows us, at a basic level, why complexity is so easy for nature to produce, and so widespread. It shows us that there are fundamental computational limitations to traditional mathematical science. It gives us insight into how similar phenomena like intelligence are to natural processes. It shows us how special—and in many ways arbitrary—the formal systems like mathematics that we have built are" (p. 178).

Wolfram concludes: "Of one thing I am certain: in the computational universe there is a huge amount that we can mine for human purposes—for creating technology, or art or other things that we as humans use" (p. 179).

The above account gives just a few glimpses from this book abundant in new ideas and insights. As the book is a collection of answers to five questions given by seventeen leading figures of the field those are really cherries picked from their research. The context is only discernible in the background, and there is an implicit hope that the reader will find the ideas interesting and read the original work of the authors. It is beyond the scope of this review to try to make any comparative analysis of different author's answers presenting such a rich and heterogeneous material. The book is a collection of brilliant thoughts from outstanding people that I find uniquely interesting and inspirational and think that many of my colleagues and students will find that too.

In sum: This is a highly recommendable and really enjoyable reading.

Craft, D. (2009). *The Crack in the Universe* (detail). Microphoto: Sandstone Rock Thin Section; 34 x 21 cm.

Cybernetics and Human Knowing. Vol. 16, nos. 3-4, pp. 201-203

Book Review
Computation, Information, Cognition:
The Nexus and the Liminal

Vincent C. Müller [1]

Computation, Information, Cognition: The Nexus and the Liminal. Edited by Susan Stuart and Gordana Dodig-Crnkovic. Cambridge Scholars Publishing, Newcastle, 2007, 340pp. ISBN: 9781847180902, £39.99, $79.99.

Are you a computer? Is your cat a computer? A single biological cell in your stomach, perhaps? And your desk? You don't think so? Well, the authors of this book suggest that you think again. They propose a computational turn, a turn towards computational explanation and towards the explanation of computation itself. The explanation of computation is the core of the present volume, but the computational turn to regard a wide variety of systems as computational is a potentially very wide-ranging project.

We have had computational machines at least since Pascal's and Leibniz' calculators, perhaps even since 1st Century BC, when the *antikithera mechanism* was built to predict lunar eclipses. However, it is only since the invention of the universal digital computer in the early 1940s and especially the spread of the Personal Computer in the 1980s that these machines have become commonplace and that their mechanism, computation, has gained more general attention beyond the confines of mathematical theory. For example, it is presently the standard assumption of cognitive science that the human mind is an computational information processor, an assumption shared by the proponents of artificial intelligence who hope that artificial computers may achieve the dizzying heights of human intelligence. Having said that, it is essential to the comprehension of the present volume that computation is not necessarily something that takes place inside certain machines that we call computers, and not even what a mathematician is doing with a pencil on a piece of paper. Computation is rather thought to be a mechanism that can be found in many systems, natural and artificial—or at least usefully postulated in the analysis and explanation of these systems. This approach has its roots in physics, but it has spread to the understanding of living systems (e.g., cells), partially incorporating earlier research programs like cybernetics and semiotics. In this sense, the presumed computation in the human mind is just one sub-system in a larger computational universe.

The papers of this volume have grown out of selected presentations at the European Computing and Philosophy conference, E-CAP, held at Mälardalen University, Sweden (http://www.idt.mdh.se/ECAP-2005/). Those papers not selected

1. Anatolia College / American College of Thessaloniki, P.O. Box 21021, 55510 Pylaia, Greece. Email: vmueller@act.edu Website: http://www.typos.de

for this volume have been published in the web journal *TripleC*, issue ii, November 2006 (http://triplec.uti.at/). The event was the third in an annual sequence of E-CAP conferences, organized each year by the European Organisation for Computing and Philosophy, which in turn is part of the International Organisation for Computing and Philosophy—an organisation that also organizes annual conferences in the US (NA-CAP) and in Asia (AP-CAP) (see http://ia-cap.org/). Volumes with selected papers from E-CAP 2007 and AP-CAP 2007 are also forthcoming. In 2008, the E-CAP meeting will take place in Montpellier in June and the NA-CAP meeting in July in Bloomington, Indiana.

Before we take a closer look at the the excellent content, allow me to mention that it is a great pity that Cambridge Scholars Publishing have done a shoddy job with the volume—or rather no job at all. The book has evidently never seen the care of a copy editor or anyone else who knows the basics of typesetting. Apart from that, the paper and binding are cheap and the printing on the dust jacket rubs off after carrying the book around in a bag for a little while—though the jacket is nicely embellished by a painting of one of the editors, Dodig-Crnkovic. Last but not least, I am told that Cambridge Scholars Publishing did not provide a copy of the finished book to each contributor—though they did provide one to this reviewer.

Evidently, in a selection such as this, there is no overall argument or thesis that can be discussed here. I will try to give a flavor of the papers and then see whether one can detect an overall direction of the exercise, perhaps even indicate some evaluation of that direction.

As the editors point out in their useful introduction, the aims of the Computing and Philosophy conferences have shifted somewhat from the use of computers in the teaching of philosophy to many other philosophical issues that arise in connection with computers, from computer modeling of the mind, to artificial intelligence and various ethical challenges posed by the growing relevance of computers in human daily life. The editors provide one-paragraph summaries of each paper in their introduction, and they organize the papers under six headings: 1. Information, 2. Ontology, 3. Bioinformation and Biosemantics, 4. Cognitive Science and Philosophy, 5. Computational Linguistics, and 6. Ethics and Education.

As mentioned above, one radical attempt at broadening the import of analysis in computational terms is to say that everything is somehow computation. One of the keynote speakers, Gregory Chatin, calls this a *digital philosophy* or *digital physics*, a neo-Pythagorean theory which says that "everything is made out of 0/1 bits, everything is digital software, and God is a computer programmer, not a mathematician!" (p. 3). This digital philosophy is proposed prominently by Edward Fredkin and Stephen Wolfram (in his book *A New Kind of Science*) and developed in Chatin's book *Meta Math!*, of which this paper is an overview. This suggestion that the world is at bottom digital is interestingly contrasted with another paper, by Pietarinen, who proposes that we need a logic of iconic (i.e. non-digital) representation. The papers by Floridi and Allo attempt to capture both digital and analog representations under the notion of *information* and to develop a (pluralistic) logic of its dynamics. These papers deal with what I take to be a central question, if not the central question: "What is computing?" and they suggest that this notion must be understood much more widely than traditionally assumed. The traditional notion described by Turing with the help of what is now called the *Turing machine* restricts computing to algorithmic processes that proceed step by step to a particular final state, taken as output. Here, computing may not be algorithmic, or even digital; in particular it will concern entities other than mere symbols.

The papers in section 2, "Ontology" seem somewhat old-fashioned in this context, in their focus on ontologies within computational systems in the conventional sense. The computationalist programme is much more at home in sections 3 to 5 where biological, cognitive and linguistic systems, respectively, are analyzed as computational ones—or where this analysis is challenged (e.g., by Miłkowski). This analysis has been commonplace in the cognitive sciences for a long time, at least in the sense of computation as information processing (and assuming that all such processing is computational). Without the cognitive baggage it has a long and successful history in the rule-fixated discipline of linguistics and it is interesting to see that it can be transferred with relative ease to biology where the notion of information appears especially useful in the explanation of functionally directed processes. Perhaps I am a computer in more than one sense, after all.

Section 6 on "Ethics and Education" is really *practical philosophy*, and even though its importance is evident (in fact more evident than that of the other directions), it is really a quite separate concern. No wonder that the impact and use of computers for teaching and ethics has inspired separate conferences, associations, and so forth.

As indicated, I see significant potential for the computational analysis of various phenomena, but if the movement wants to be one of the kind that propose an analysis that "Everything is x, at bottom," then we need to know more about that x. The widening of computation often goes together with a rejection of Turing's notion of computation. Fine, but if we do not want that old-fashioned notion, what will it be?

It emerges that these papers offer much more than just *something about philosophy and computing*, indeed they went well beyond a philosophy of computation (or, worse, a philosophy of computers) into an overarching research programme of a computational philosophy of the world. This programme is a promising one and how far reaching it will be cannot be seen at the moment. Indeed, I would expect that computational analyses are on their way toward becoming mainstream and will soon be so common that we will hardly notice them at all.

Craft, D. (2009). *Broken Symmetries*. Photomontage: Triple Golden Rectangles; 21 x 102 cm.

Craft, D. (2009). *Angular Momentum*. Microphoto: Benzoic Acid Melt Crystals; 34 x 21 cm.